T0275760

CAMBRIDGE STUDIES IN
ADVANCED MATHEMATICS 28

Topics in metric fixed point theory

Already published

Topics in metric fixed point theory

KAZIMIERZ GOEBEL

Institute of Mathematics
Maria Curie Sklodowska University, Lublin, Poland

W. A. KIRK

Department of Mathematics
University of Iowa, Iowa City, Iowa, USA

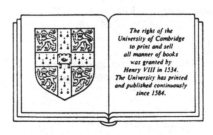

The right of the
University of Cambridge
to print and sell
all manner of books
was granted by
Henry VIII in 1534.
The University has printed
and published continuously
since 1584.

CAMBRIDGE UNIVERSITY PRESS

Cambridge
New York   Port Chester   Melbourne   Sydney

CAMBRIDGE UNIVERSITY PRESS
Cambridge, New York, Melbourne, Madrid, Cape Town, Singapore, São Paulo

Cambridge University Press
The Edinburgh Building, Cambridge CB2 8RU, UK

Published in the United States of America by Cambridge University Press, New York

www.cambridge.org
Information on this title: www.cambridge.org/9780521382892

First published 1990
This digitally printed version 2008

A catalogue record for this publication is available from the British Library

Library of Congress Cataloguing in Publication data

Goebel, Kazimierz, 1940–
Topics in metric fixed point theory/Kazimierz Goebel and W. A. Kirk.
p. cm. – (Cambridge studies in advanced mathematics)
Includes bibliographical references.

1. Fixed point theory.   I. Kirk, W. A.   II. Title.   III. Series.
QA329.9.G64   1990
515′.7248—dc20   89-70858   CIP

ISBN 978-0-521-38289-2 hardback
ISBN 978-0-521-06406-4 paperback

Contents

Preface

The term 'Metric' Fixed Point Theory refers to those fixed point theoretic results in which geometric conditions on the underlying spaces and/or mappings play a crucial role. Obviously there can be no clear line separating this branch of fixed point theory from either the topological or set-theoretic branches since metric methods are often useful in proving results which are basically nonmetric in nature, and vice versa. However, the results considered here are always couched in at least a metric space framework, usually in a Banach space setting, and the methods typically involve both the topological and the geometric structure of the space in conjunction with metric constraints on the behavior of the mappings.

For the past twenty-five years metric fixed point theory has been a flourishing area of research for many mathematicians. Although a substantial number of definitive results have now been discovered, a few questions lying at the heart of the theory remain open and there are many unanswered questions regarding the limits to which the theory may be extended. Some of these questions are merely tantalizing while others suggest substantial new avenues of research.

It is apparent that the theory has now reached a level of maturity appropriate to an examination of its central themes. The topics selected for this text were chosen accordingly. No attempt has been made to explore all aspects of the theory nor to present a compendium of known facts. Our objective is merely to offer the mathematical community an accessible self-contained document which can be used as an introduction to the subject and its development. We have attempted to render the major results understandable to a wide audience, including nonspecialists, and at the same time to provide a source for examples, references, open questions, and (occasionally) new approaches for those currently working in the subject. The results presented in detail were selected to illustrate the directions research in this field has taken during the past twenty-five years and to illustrate the flavor of the subject. Also, in our attempt to render the topics treated self-contained, we only assume familiarity with the basic concepts of analysis and topology. To the extent these goals have been achieved, the text should be of interest to graduate students seeking a field of interest, to mathematicians interested in learning about the subject, and to specialists.

The structure of the text is straightforward. There are twenty-one short chapters devoted to various aspects of the theory. A substantial portion of the book, Chapters 3–13, is devoted to the classical theory of nonexpansive mappings. Included in these chapters is a discussion of the basic problems of the field: the existence of fixed points, the structure of the fixed point sets, and approximation techniques for locating fixed points. Chapter 14, Ultra-filter Methods, is exceptional in that it contains some recent results obtained by utilizing 'nonstandard' techniques based on the concepts of ultrafilters, ultrapowers and ultranets. These methods are nonintuitive and not usually viewed as 'metric geometry' tools. Nevertheless, they are powerful techniques which seem capable of laying the foundation of an entirely separate branch of the subject. Since results in this direction are still emerging, we provide only an introduction to the subject here.

Chapters 14, 15 and 16 are devoted to some generalizations and extensions of the previous results to classes of mappings which are not necessarily nonexpansive but which satisfy closely related metric constraints, and the last four chapters contain some relatively fresh problems of metrical type which evolved from the classical theorems of Brouwer and Schauder (or, more precisely, from 'naive' attempts to obtain generalizations of these results to noncompact settings).

The theory treated here has many contributors. Those who developed the classical theory include the celebrated mathematicians L. E. J. Brouwer, S. Banach, and J. Schauder. The metric theory was given a new impetus by the 1965 fixed point theorems for nonexpansive mappings discovered independently by F. Browder, D. Göhde, and W. A. Kirk, and by the widely circulated *1967 Lecture Notes* of Z. Opial. There have been numerous major discoveries since then, and this text is made possible only because of those contributions as well as the contributions of many pioneering mathematicians who developed the functional analytic framework within which most of the subject is couched.

We thank collectively our many friends and colleagues who, through their encouragement and help, influenced the development of this book. We are particularly grateful to Juan Gatica, who examined portions of the manuscript in detail, and to Stanislaw Prus, whose astute observations led to significant improvements in Chapter 9. And we especially thank Tadeusz Kuczumow for pointing out numerous oversights in the original draft of the manuscript. Finally, we thank our typist, Julie Hill, for skilfully and patiently seeing the manuscript through its various stages.

1

Preliminaries

To facilitate readability of the text we shall frequently recall basic definitions and facts as the need arises. Nevertheless, certain topics, such as those related to reflexivity of Banach spaces and the weak and weak* topologies, play such a primary role in our development that they deserve special attention. It is our purpose in this chapter to collect facts which will be used repeatedly throughout the text.

For the most part we use notation and symbols commonly used in textbooks. We only mention that the symbols $\mathbb{N}$, $\mathbb{Z}$, $\mathbb{R}$, $\mathbb{C}$ are reserved, respectively, for the natural numbers, integers, real numbers, and complex numbers.

If A is a subset of a metric space (M, ρ) and if $x \in M$, then diam A and dist(x, A) denote, respectively, the diameter of A and the distance from x to A. Precisely,

$$\text{diam } A = \sup\{\rho(x, y): x, y \in A\};$$
$$\text{dist}(x, A) = \inf\{\rho(x, y): y \in A\}.$$

Also $B(x; r)$ always denotes the *closed* ball centered at x with radius $r > 0$:

$$B(x; r) = \{y \in M : \rho(x, y) \leqslant r\}.$$

We use the symbol $\bar{A}$ to denote the closure of A in M, and if M has a topology τ other than the one induced by the metric, we use ${}^{\tau}\bar{A}$ to denote the closure of A in (M, τ).

For the remainder of the chapter we confine our attention to Banach spaces. All the above definitions carry over to a Banach space setting $(X, \| \ \|)$ by taking $\rho(x, y) = \|x - y\|$.

We remark at the outset that while most of the results presented in this book remain valid for complex Banach spaces, we shall, for convenience, confine our attention to real Banach spaces. Throughout the remainder of the chapter $X = (X, \| \ \|)$ will denote an arbitrary real Banach space.

Convexity We use conv A to denote the *convex hull* of $A \subset X$, i.e., the smallest convex subset of X which contains A. Obviously

$$\text{conv } A = \cap\{K \subset X : K \supset A, K \text{ is convex}\}.$$

1

Moreover, $x \in \operatorname{conv} A$ if and only if x is of the form $x = \Sigma_{i=1}^{n} \lambda_i x_i$ where $x_i \in A$, $\lambda_i \geqslant 0$ and $\Sigma_{i=1}^{n} \lambda_i = 1$. The closure of $\operatorname{conv} A$ is denoted $\overline{\operatorname{conv}} A$ and called the *convex closure* of A. As above,

$$\overline{\operatorname{conv}} A = \cap\{K \subset X : K \supset A, K \text{ is closed and convex}\}.$$

A fundamental property of the convex closure operation is given by the following:

Theorem 1.1 (Mazur's) *If $\bar{A}$ is compact, then so is $\overline{\operatorname{conv}} A$.*

Dual spaces and reflexivity For two Banach spaces X and Y, let $\mathscr{L}(X, Y)$ denote the space of all bounded (continuous) linear operators (mappings) from X to Y. The norm $\|T\|$ of an operator $T \in \mathscr{L}(X, Y)$ is given by

$$\|T\| = \sup\{\|Tx\|/\|x\| : x \in X, x \neq 0\}$$
$$= \sup\{\|Tx\| : x \in X, \|x\| = 1\}.$$

It is easily verified that if X and Y are Banach spaces then so is $(\mathscr{L}(X, Y), \| \ \|)$. The *dual* or *conjugate space* X^* of X is the space $X^* = \mathscr{L}(X, \mathbb{R})$. The elements of X^* are called *continuous linear functionals*. For $x^* \in X^*$, we shall often use the notation which pairs elements of X with elements of X^*:

$$x^*(x) = \langle x, x^* \rangle, x \in X, x^* \in X^*.$$

(This of course is consistent with the pairing in a Hilbert space H where, by the Riesz Representation Theorem, $H = H^*$ and the value of $y(x)$ for $x \in H$, $y \in H^*$ is given by the usual inner product $\langle x, y \rangle$.)

The space $X^{**} = \mathscr{L}(X^*, \mathbb{R})$ is called the second dual (or conjugate) space of X. If $x \in X$ is fixed, then the relation $\langle x, x^* \rangle$ defines a continuous linear functional on X^*; thus x is associated in a natural way with an element x^{**} of X^{**}. The mapping $x \mapsto x^{**}$ is called the *canonical* (or *natural*) embedding of X in X^{**}. This embedding is always a linear isometry. If it is also surjective then X is said to be *reflexive* and we write $X = X^{**}$.

The weak and weak* topologies The *weak topology on* X is the topology generated by the family of seminorms $\{p_{x^*}\}$, $x^* \in X^*$, where

$$p_{x^*}(x) = |\langle x, x^* \rangle|, \qquad x \in X.$$

Similarly, the *weak* topology on* X^* is generated by the seminorms $\{p_x\}$, $x \in X$, where

$$p_x(x) = |\langle x, x^* \rangle|, \qquad x^* \in X^*.$$

Both X and X^* are locally convex, linear topological spaces relative to their respective weak and weak* topologies. Note also that X^* also has a weak topology which in general is distinct from its weak* topology. The two, of course, coincide when X is reflexive.

The weak topology on X is the weakest (coarsest) topology for which all the functionals $x^* \in X^*$ are continuous. In particular, a net $\{x_\alpha, \alpha \in A\}$ converges to an element $x \in X$ in the weak topology if and only if $\lim_\alpha \langle x_\alpha, x^* \rangle = \langle x, x^* \rangle$ for each $x^* \in X^*$. When this occurs we say that $\{x_\alpha\}$ is *weakly convergent* or *converges weakly* to x, and we write

$$w\text{-}\lim_\alpha x_\alpha = x.$$

Similarly, a net $\{x_\alpha^*, \alpha \in A\}$ in X^* converges to $x^* \in X^*$ in the weak* topology if and only if for each $x \in X$, $\lim_\alpha \langle x, x_\alpha^* \rangle = \langle x, x^* \rangle$, in which case

$$w^*\text{-}\lim_\alpha x_\alpha^* = x^*.$$

(We elaborate on the notion of net convergence in Chapter 14.)

We now collect some basic and well-known properties about the weak and weak* topologies.

Property 1.1 *A convex subset K of X is closed if and only if it is weakly closed.*

Property 1.2 *If K is a weakly compact subset of X then $\overline{\text{conv}}\, K$ is also weakly compact.*

The above facts do not carry over to the weak* topology. However the following fact about the weak* topology is very important.

Property 1.3 (Alaoglu's Theorem) *The unit ball $B(0; 1)$ in a dual space X^* is always compact in the weak* topology.*

Note that the above implies that any ball or any intersection of balls in a dual space is weak* compact.

If X is reflexive, then $X = X^{**}$. Thus, in view of Alaoglu's Theorem, we have:

Property 1.4 *If X is reflexive, then each ball in X is compact in the weak topology.*

The above, in fact, characterizes reflexive spaces: A Banach space X is reflexive if and only if its unit ball is compact in the weak topology. When combined with Properties 1.1 and 1.2 this fact implies that each bounded, closed and convex subset of a reflexive space is compact in the weak topology (i.e., weakly compact).

The following rather deep fact is of fundamental importance. We invoke this fact repeatedly and without comment.

Property 1.5 (*The Eberlein–Smulian Theorem*) *For any subset A of X the following statements are equivalent.*

(a)   *Each sequence $\{x_n\}$ in A has a subsequence that is weakly convergent.*
(b)   *Each sequence $\{x_n\}$ in A has a weak cluster point in X.*
(c)   *The weak closure $^w\bar{A}$ of A is weakly compact.*

Thus weak compactness is always equivalent to sequential weak compactness. This fact does not hold for the weak* topology. However the following is true.

Property 1.6 (*The Krein–Smulian Theorem*) *A subset K of a dual space X^* is weak* closed if and only if for each $r>0$ the sets $\{x^* \in K: \|x^*\| \leqslant r\}$ are weak* closed.*

The fact that the weak* topology on a separable space is metrizable gives rise to the following useful fact.

Property 1.7 *If X is separable and if K is a convex subset of X^*, then K is weak* closed if and only if K is weak* sequentially closed.*

Proofs of the above facts may be found, for example, in Dunford and Schwartz (1957) or in almost any standard text in functional analysis. For our purposes, however, the mere knowledge of these facts will enable the reader to follow all the steps in the proofs which follow.

We will now list several properties which characterize reflexivity. Other such characterizations will be introduced in the text as needed.

Property 1.8 *A Banach space X is reflexive if and only if one of the following (equivalent) conditions holds.*

(a)   *X^* is reflexive.*
(b)   *$B(0; 1)$ is weakly compact in X^*.*
(c)   *Any bounded sequence in X has a weakly convergent subsequence.*

(d) (James (1957)). *For any $x^* \in X^*$ there exists $x \in B(0; 1)$ such that $x^*(x) = \|x^*\|$.*

(e) (James (1964)). *For any bounded, closed and convex subset K of X and any $x^* \in X^*$ there exists $x \in K$ such that $x^*(x) = \sup\{x^*(y): y \in K\}$.*

(f) (Smulian (1939)). *For any decreasing sequence $\{K_n\}$ of nonempty, bounded, closed and convex subsets of X, $\bigcap_{n=1}^{\infty} K_n \neq \emptyset$.*

There is one important fact that one should keep in mind when dealing with the weak* topology on a dual space X^*. Namely, two different Banach spaces, X_1 and X_2 may have the same dual space: $X^* = X_1^* = X_2^*$ (where the dual norms on X_1^* and X_2^* coincide). However the weak* topologies induced on X^* by X_1 and X_2 may differ. This occurs, for example, with the classical space l^1 which is dual to both c (the space of convergent real sequences) and c_0 (the space of such sequences which converge to 0).

We also assume the reader is familiar with the classical Banach spaces which arise frequently in the literature (e.g., the l^p- and L^p- spaces, $1 \leqslant p \leqslant \infty$, $\mathscr{C}[0, 1]$, c, c_0, etc.) along with the facts: $(L^p)^*$ and $(l^p)^*$ $(1 < p < \infty)$ are represented, respectively, by L^q and l^q where $p^{-1} + q^{-1} = 1$, $(l^1)^*$ is represented by l^{∞}, etc.

Finally, we shall require some basic facts about Schauder bases. A sequence $\{e^n\}$ in a Banach space X is called a *Schauder basis* for X if for any $x \in X$ there is a unique sequence $\{\xi_n\} = \{\xi_n(x)\}$ of real numbers such that

$$x = \sum_{n=1}^{\infty} \xi_n e^n = \lim_{k \to \infty} \sum_{n=1}^{k} \xi_n e^n.$$

Obviously if X has such a basis then all the e^n are linearly independent. Also there exist two sequences $\{P_k\}$ and $\{R_k\}$ of natural operations; the projections:

$$P_k: x \mapsto \sum_{n=1}^{k} \xi_n e^n;$$

and the remainders:

$$R_k: x \mapsto \sum_{n=k+1}^{k} \xi_n e^n.$$

The family $\{P_k\}$ (as well as $\{R_k\}$) is equicontinuous (Banach-Steinhaus Theorem) and the constant $K = \sup_n \|P_n\|$ is called the basis constant for X. Moreover, for any $x \in X$, $\lim_{n \to \infty} P_n x = x$; $\lim_{n \to \infty} R_n x = 0$. Consequently, any linear functional $f \in X^*$ has the following representation:

$$fx = f\left(\sum_{n=1}^{\infty} \xi_n e^n\right) = \sum_{n=1}^{\infty} \xi_n f(e^n).$$

Thus f is completely determined by its values at the basis elements. In particular, the coordinate functionals $\{f_n\}$ defined by the relation $f_n(e^i) = \delta_{in}$ are *biorthogonal* to the basis $\{e^n\}$. The family $\{f_n\}$ is equibounded, and the value $\xi_n = f_n(x)$ is called the nth coordinate of x with respect to the basis $\{e^n\}$. We shall call a sequence $\{x_n\}$ in X *coordinate-wise convergent* if for each $n \in \mathbb{N}$, the sequence $\{f_n(x_k)\}_{k=1}^{\infty}$ converges. (In general this type of convergence is not equivalent to either weak or weak* convergence.)

The sequence $\{f_n\}$ of coordinate functionals is not necessarily a basis for the space X^*. However this is the case if the basis $\{e^n\}$ has the following property: For any $f \in X^*$ the norm of f restricted to span $\{e^k, e^{k+1}, \ldots\}$ converges to 0 as $k \mapsto \infty$. Bases which have this property are said to be *shrinking*. Also, a basis $\{e^n\}$ of X is said to be *boundedly complete* if for every sequence $\{\xi_n\}$ of scalars for which $\sup_k \|\Sigma_{n=1}^k \xi_n e^n\| < \infty$, the series $\Sigma_{n=1}^{\infty} \xi_n e^n$ converges, and thus represents an element of X. These concepts provide a connection between basis theory and reflexivity via the following result.

Property 1.9 (James (1950)). *A Banach space X with a basis $\{e^n\}$ is reflexive if and only if $\{e^n\}$ is shrinking and boundedly complete.*

In view of Property 1.9 a sequence $\{x_k\}$ of elements of a reflexive Banach space which has a basis converges weakly if and only if it is bounded and coordinate-wise convergent.

We shall use the above facts mostly in special settings, for example, in such spaces as c_0, l^1, l^p ($1 < p < \infty$) with the natural basis $\{e^n\} = \{\delta_{in}\}$. We note, in particular, that $\{e^n\}$ is shrinking but *not* boundedly complete in c_0 and l^1.

For those interested in proof of most of the above facts, we suggest the book by van Dulst (1978) which is devoted exclusively to reflexive spaces. Also, persons interested in knowing more about the geometry of Banach spaces might wish to consult such books as Day (1973), Lindenstrauss and Tzafriri (1977, 1979) or Diestel (1975).

2

Banach's Contraction Principle

The fixed point theorem, generally known as the Banach Contraction Principle, appeared in explicit form in Banach's thesis in 1922 where it was used to establish the existence of a solution for an integral equation. Since then, because of its simplicity and usefulness, it has become a very popular tool in solving existence problems in many branches of mathematical analysis. In this chapter we prove Banach's Contraction Principle, discuss some of its more useful variants, and present a few diverse examples of its applications.

Let M be a metric space with distance function (metric) ρ. A mapping $T: M \to M$ is said to be *lipschitzian* if there exists $k \geqslant 0$ such that for all $x, y \in M$,

$$\rho(Tx, Ty) \leqslant k\rho(x, y). \tag{2.1}$$

The smallest k for which (2.1) holds is said to be the *Lipschitz constant* for T. We shall often denote the respective Lipschitz constants of different mappings T and S by $k(T)$ and $k(S)$ and when relevant $k_\rho(T)$ will be used to denote the Lipschitz constant of T with respect to the metric ρ.

For two mappings $S, T: M \to M$,

$$k(T \circ S) \leqslant k(T)k(S)$$

and, in particular,

$$k(T^n) \leqslant k^n(T), \qquad n = 1, 2, \ldots.$$

If M is a linear space whose metric is generated by a norm, $k(T+S) \leqslant k(T) + k(S)$ and, for $\alpha \geqslant 0$, $k(\alpha T) = \alpha k(T)$.

A mapping $T: M \to M$ is said to be a *contraction* if $k(T) < 1$; more precisely, T is a k-contraction with respect to ρ if $k_\rho(T) \leqslant k < 1$.

Theorem 2.1 (Banach's Contraction Principle) *Let (M, ρ) be a complete metric space and let $T: M \to M$ be a contraction. Then T has a unique fixed point in M, and for each $x_0 \in M$ the sequence of iterates $\{T^n x_0\}$ converges to this fixed point.*

We give three proofs of Theorem 2.1. The first is nonconstructive and establishes only the existence part of the theorem while the second, which is a variant of the original proof, not only provides the existence of a fixed point but, as in the original proof, also provides a method for its approximation. We then give the original (and commonly known) proof.

Proof 1 Let $a=\inf\{\rho(x, Tx): x \in M\}$ and $k=k_\rho(T)$. To see that $a=0$, let $\epsilon>0$ and select $x \in M$ so that $\rho(x, Tx) \leqslant a+\epsilon$. Then

$$a \leqslant \rho(Tx, T^2 x) \leqslant k\rho(x, Tx) \leqslant k(a+\epsilon).$$

Since $k<1$ and ϵ can be taken arbitrarily small, $a=0$.

Now for any $\epsilon>0$ the set

$$M_\epsilon = \{x \in M : \rho(x, Tx) \leqslant \epsilon\}$$

is nonempty and closed. Moreover, for any $x, y \in M_\epsilon$,

$$\rho(x, y) \leqslant \rho(x, Tx) + \rho(Tx, Ty) + \rho(Ty, y) \leqslant 2\epsilon + k\rho(x, y),$$

yielding

$$\rho(x, y) \leqslant \frac{2\epsilon}{1-k}$$

from which $\lim_{\epsilon \to 0} \operatorname{diam} M_\epsilon = 0$.

Since the family $\{M_\epsilon\}$ descends as $\epsilon \downarrow 0$, the Cantor Intersection Theorem implies $\bigcap_{\epsilon>0} M_\epsilon$ consists of exactly one point, say x, which must be fixed under T ($x = Tx$).

Proof 2 Set $\varphi(x)=(1-k)^{-1}\rho(x, Tx)$ for $x \in M$ (where $k=k_\rho(T)$). Then

$$\rho(x, Tx) - k\rho(x, Tx) \leqslant \rho(x, Tx) - \rho(Tx, T^2 x);$$

hence

$$\rho(x,Tx) \leqslant \varphi(x) - \varphi(Tx), \qquad x \in M. \tag{2.2}$$

Thus for $x_0 \in M$ and $n, m \in \mathbb{N}$ with $n<m$,

$$\rho(T^n x_0, T^{m+1} x_0) \leqslant \sum_{i=n}^{m} \rho(T^i x_0, T^{i+1} x_0) \leqslant \varphi(T^n x_0) - \varphi(T^{m+1} x_0). \tag{2.3}$$

In particular, $\Sigma_{i=0}^{\infty} \rho(T^i x_0, T^{i+1} x_0) < +\infty$. Therefore $\{T^n x_0\}$ is a Cauchy sequence and, since T is continuous, it converges to a fixed point x of T. The rate of this convergence may be obtained from (2.3) by letting $m \to \infty$:

$$\rho(T^n x_0, x) \leqslant \varphi(T^n x_0) = (1-k)^{-1} \rho(T^n x_0, T^{n+1} x_0) \leqslant \frac{k^n}{1-k}\rho(x_0, Tx_0).$$

Remark 2.1 The above proof shows that *any* continuous mapping which satisfies (2.2) for arbitrary $\varphi: M \to \mathbb{R}^+$ must have a fixed point. In fact, it can be shown by other means that if φ is lower semicontinuous, then an *arbitrary* mapping $T: M \to M$ satisfying (2.2) must have a fixed point. This fact, which is generally known as the Caristi Theorem, is presented in detail later. It is equivalent to the Ekeland Minimization Principle (Ekeland, 1974) (assuming the Axiom of choice) and has many applications in analysis (see, e.g., Brezis and Browder, 1976, for a thorough discussion). The fixed point in both the above cases need not be unique and in the second instance the sequence $\{T^n x_0\}$ need not even converge to a fixed point of T.

Proof 3 Select $x_0 \in M$ and define the iterative sequence $\{x_n\}$ by $x_{n+1} = T x_n$ (equivalently, $x_n = T^n x_0$), $n = 0, 1, 2, \ldots$. Observe that for any indices $n, p \in \mathbb{N}$,

$$
\begin{aligned}
\rho(x_n, x_{n+p}) &= \rho(T^n x_0, T^{n+p} x_0) = \rho(T^n x_0, T^n \circ T^p x_0) \leqslant k(T^n)\rho(x_0, T^p x_0) \\
&\leqslant k^n [\rho(x_0, Tx_0) + \rho(Tx_0, T^2 x_0) + \cdots + \rho(T^{p-1} x_0, T^p x_0)] \\
&\leqslant k^n (1 + k + \cdots + k^{p-1})\rho(x_0, Tx_0) \\
&\leqslant k^n \left(\frac{1-k^p}{1-k}\right)\rho(x_0, Tx_0).
\end{aligned}
\tag{2.4}
$$

This shows that $\{x_n\}$ is a Cauchy sequence, and since M is complete there exists $x \in M$ such that $\lim_{n \to \infty} x_n = x$. To see that x is the unique fixed point of T, observe that

$$
x = \lim_{n \to \infty} x_n = \lim_{n \to \infty} x_{n+1} = \lim_{n \to \infty} Tx_n = Tx
$$

and, moreover, $x = Tx$ and $y = Ty$ imply

$$
\rho(x, y) = \rho(Tx, Ty) \leqslant k\rho(x, y),
$$

yielding $\rho(x, y) = 0$.

As in the second proof, letting $p \to \infty$ in (2.4) yields

$$
\rho(x_n, x) = \rho(T^n x_0, x) \leqslant \frac{k^n}{1-k}\rho(x_0, Tx_0).
\tag{2.5}
$$

Remark 2.2 An analysis of the third proof reveals that the assumption $k(T) < 1$ is stronger than necessary. It suffices to assume $k(T^n) < 1$ for at least one fixed $n \in \mathbb{N}$. This implies T^n is a contraction and (by Theorem 2.1) has a unique fixed point x. But $Tx = T^{n+1}x = T^n \circ Tx$, so Tx is also a fixed point of T^n. Hence $x = Tx$, proving x is also a fixed point of T (and the unique one). It is not difficult to find examples of mappings T (even on the interval $[0, 1]$) which are continuous (or discontinuous) and for which $k(T^n) < 1$ while $k(T) \geqslant 1$. However, these examples seem exceptional, so we shall confine our attention to more typical situations.

We now expand on the idea of Remark 2.2. Let $T: M \rightarrow M$ be lipschitzian, fix $x_0 \in M$, and let $x_n = T^n x_0$. The counterpart of the estimate (1.2) for this more general class of mappings is

$$\rho(x_n, x_{n+p}) \leqslant \sum_{i=n}^{n+p-1} \rho(T^i x_0, T^{i+1} x_0)$$

$$\leqslant \left[\sum_{i=0}^{p} k(T^{n+i}) \right] \rho(x_0, Tx_0).$$

Thus $\{x_n\}$ is a Cauchy sequence if it is the case that

$$\sum_{i=1}^{\infty} k(T^i) < +\infty. \tag{2.6}$$

Using the fact that $k(T^n)$ is multiplicative, i.e., since $k(T^{n+m}) \leqslant k(T^n)k(T^m)$, it is easy to see that there exists a number $k_\infty(T)$ which satisfies:

$$k_\infty(T) = \lim_{n \to \infty} [k(T^n)]^{1/n} = \inf\{[k(T^n)]^{1/n} : n = 1, 2, \dots\}. \tag{2.7}$$

Thus (2.6) holds if and only if $k_\infty(T) < 1$, so the assumption $k(T) < 1$ in Theorem 2.1 can be replaced with $k_\infty(T) < 1$.

Of course the question remains of whether the weaker assumption $k_\infty(T) < 1$ actually provides a stronger version of Theorem 2.1. To respond to this, we introduce the notion of equivalence between metrics: Two metrics ρ and r defined on a given set M are said to be equivalent if there exist two positive constants a and b such that for all $x, y \in M$,

$$ar(x, y) \leqslant \rho(x, y) \leqslant br(x, y). \tag{2.8}$$

For two such metrics, any sequence which is Cauchy with respect to r is also Cauchy with respect to ρ (and conversely). Consequently, (M, ρ) is complete if and only if (M, r) is also.

For a ρ-lipschitzian mapping $T: M \rightarrow M$, (2.8) implies

$$r(Tx, Ty) \leqslant (1/a)\rho(Tx, Ty) \leqslant (1/a)k_\rho(T)\rho(x, y) \leqslant (b/a)k_\rho(T)r(x, y),$$

and thus $k_r(T) \leqslant (b/a)k_\rho(T)$. Similarly, $k_\rho(T) \leqslant (b/a)k_r(T)$, and so for any $n \in \mathbb{N}$,

$$\frac{a}{b} k_\rho(T^n) \leqslant k_r(T^n) \leqslant \frac{b}{a} k_\rho(T^n).$$

Consequently,

$$\lim_{n\to\infty} [k_r(T^n)]^{1/n} = \lim_{n\to\infty} [k_\rho(T^n)]^{1/n},$$

and this shows that $k_\infty(T)$ is constant with respect to equivalent metrics. Moreover, in view of (2.7), $k_\infty(T) \leqslant k_r(T)$ for all metrics r equivalent to ρ. On the other hand, for any $\lambda \in [0, 1/k_\infty(T))$ the series

$$r_\lambda(x, y) = \sum_{n=0}^{\infty} \lambda^n \rho(T^n x, T^n y)$$

converges and yields a metric, r_λ, equivalent to ρ:

$$\rho(x, y) \leqslant r_\lambda(x, y) \leqslant \left[\sum_{n=0}^{\infty} k_\rho(T^n)\lambda^n \right] \rho(x, y).$$

Thus

$$r_\lambda(Tx, Ty) = \sum_{n=0}^{\infty} \lambda^n \rho(T^{n+1}x, T^{n+1}y)$$

$$= \frac{1}{\lambda}[r_\lambda(x, y) - \rho(x, y)]$$

$$\leqslant \frac{1}{\lambda} r_\lambda(x, y);$$

hence $k_{r_\lambda}(T) \leqslant 1/\lambda$. Taking $\lambda = 1/[k_\infty(T)+\epsilon]$, $\epsilon > 0$, we have

$$k_{r_\lambda}(T) \leqslant k_\infty(T) + \epsilon$$

and so we conclude:

$$k_\infty(T) = \inf k_r(T),$$

where the infimum is over all metrics r equivalent to ρ.

To summarize: *Any mapping $T: M \to M$ for which $k_\infty(T) < 1$ is a contraction with respect to a properly chosen equivalent metric.* In principle, then, the assumption $k_\infty(T) < 1$ does not yield a stronger version of Theorem 2.1. However, as we shall see, the choice of a proper metric is sometimes helpful in applications because it provides nice estimates on the rate of convergence of iterates.

Perhaps the most obvious question raised by the study of contraction mappings is: What happens when $k(T) = 1$? The simple example $Tx = x + 1$ for $x \in \mathbb{R}$ shows that the counterpart of Banach's Theorem fails to hold. However, within the context of a wide class of standard spaces, the bounded, closed and convex subsets of Banach spaces, a rich fixed point theory for such mappings exist. We shall explore this in subsequent chapters.

There is a natural class of mappings which falls properly between the contraction mappings and those mappings T for which $k(T) = 1$. A mapping $T: M \to M$ is called *contractive* (or strictly contractive) if:

$$\rho(Tx, Ty) < \rho(x, y), \qquad x, y \in M, x \neq y. \qquad (2.9)$$

Obviously a mapping of this type can have at most one fixed point. The mapping $T: \mathbb{R} \to \mathbb{R}$ defined by $Tx = 1 + \ln(1 + e^x)$ provides a simple example of a fixed point free, contractive mapping. (In fact, $|x - Tx| > 1$ for all $x \in \mathbb{R}$.) However, such mappings always have fixed points in compact spaces.

Theorem 2.2 *Let (M, ρ) be a compact metric space and let $T: M \to M$ be contractive. Then T has a unique fixed point in M, and for any $x_0 \in M$ the sequence $\{T^n x_0\}$ of iterates converges to this fixed point.*

Proof The function $\varphi: M \to \mathbb{R}^+$ defined by $\varphi(y) = \rho(y, Ty)$ is continuous on M and hence by compactness attains its minimum, say at $x \in M$. If $x \neq Tx$ then $\varphi(Tx) = \rho(Tx, T^2 x) < \rho(x, Tx)$ – a contradiction. So $x = Tx$. Now let $x_0 \in M$ and set $a_n = \rho(T^n x_0, x)$. Since

$$a_{n+1} = \rho(T^{n+1} x_0, x) = \rho(T^{n+1} x_0, Tx) \leqslant \rho(T^n x_0, x) = a_n,$$

$\{a_n\}$ is a nonincreasing sequence of nonnegative real numbers and so has a limit, say a. Again by compactness, $\{T^n x_0\}$ has a convergent subsequence $\{T^{n_k} x_0\}$; say $\lim_{k \to \infty} T^{n_k} x_0 = z$. Obviously $\rho(z, x) = a$. If $a > 0$, then we obtain the contradiction:

$$a = \lim_{k \to \infty} \rho(T^{n_k+1} x_0, x) = \rho(Tz, x) = \rho(Tz, Tx) < \rho(z, x) = a.$$

Thus $a = 0$. Therefore any convergent subsequence of $\{T^n x_0\}$ must converge to x, so by compactness, $\lim_{n \to \infty} T^n x_0 = x$.

Theorem 2.2 is generally attributed to Edelstein (1962) (who actually obtained slightly more general versions).

Compactness in Theorem 2.2 cannot be replaced with completeness and boundedness:

Example 2.1 Let $\mathscr{C}[0, 1]$ be the space of continuous real valued functions on $[0, 1]$ with the standard supremum norm: For $x \in \mathscr{C}[0, 1]$,

$$\|x\| = \sup\{|x(t)|: t \in [0, 1]\}.$$

Set $M = \{x \in \mathscr{C}[0, 1]: 0 = x(0) \leqslant x(t) \leqslant x(1) = 1\}$. M is closed (also bounded and convex), and since $\mathscr{C}[0, 1]$ is complete in the metric induced by $\| \ \|$, so is M. The mapping $T: M \to M$ defined by $(Tx)(t) = tx(t)$, $x \in M$, $t \in [0, 1]$, is both contractive and fixed point free.

The simplicity and usefulness of Banach's Contraction Principle has inspired many authors to analyze it further. These studies have led to a number of generalizations and modifications of the principle, one of the deepest of which was alluded to in Remark 2.1.

Theorem 2.3 (Caristi) *Let* (M, ρ) *be a complete metric space and let* $\varphi: M \to \mathbb{R}$ *be a lower semicontinuous function which is bounded below. Suppose* $T: M \to M$ *is an arbitrary mapping which satisfies:*

$$\rho(u, T(u)) \leqslant \varphi(u) - \varphi(T(u)), \; u \in M.$$

Then T has a fixed point.

The relation of the above to the Banach Contraction Principle was noted in Remark 2.1 above. However, Caristi's result is really only a distant relative of the Banach Principle. Caristi's original proof (Caristi, 1976) and Wong's refinement (Wong, 1976) of it use transfinite induction. Several elegant proofs which use Zorn's Lemma are known (e.g., Kirk, 1976; Pasicki, 1978) and somewhat intricate constructive proofs are also known to exist (e.g., Siegel, 1977). Here we derive Caristi's theorem from the following reformulation of a result of Brezis and Browder (1976). In this theorem X denotes a partially ordered (preordered) set, and for $x \in X, S(x) = \{ y \in X : y \geqslant x \}$. A sequence $\{x_n\}$ in X is said to be increasing provided $x_n \leqslant x_{n+1}$ for all n.

Theorem 2.4 *Let* $\psi: X \to \mathbb{R}$ *be a function which satisfies:*

(a) $x \leqslant y$ *and* $x \neq y$ *implies* $\psi(x) < \psi(y)$.
(b) *for any increasing sequence* $\{x_n\}$ *in X such that* $\psi(x_n) \leqslant C < \infty$ *for all n there exists some* $y \in X$ *such that* $x_n \leqslant y$ *for all n;*
(c) *for each* $x \in X, \psi(S(x))$ *is bounded above.*
Then for each $x \in X$ *there exists* $x' \in S(x)$ *such that x' is maximal, i.e.,* $\{x'\} = S(x')$.

Proof For $a \in X$ let $\rho(a) = \sup\{\psi(b) : b \in S(a)\}$. Assume the conclusion of the theorem fails for some $x \in X$ and define by induction a sequence $\{x_n\}$ such that $x_1 = x$, and $x_{n+1} \in S(x_n)$ satisfies $\rho(x_n) \leqslant \psi(x_{n+1}) + (1/n)$ for all $n \in \mathbb{N}$. Since $\psi(x_{n+1}) \leqslant \rho(x) < \infty$, it follows from (2) that there exists some $y \in X$ such that $x_n \leqslant y$ for all n. Also, by assumption, y is not maximal in $S(x)$ so there exists $u \in X$ such that $y \leqslant u$ and $\psi(y) < \psi(u)$. Since $x_n \leqslant u, \psi(u) \leqslant \rho(x_n)$ for all n. Also, $x_{n+1} \leqslant y$; thus $\psi(x_{n+1}) \leqslant \psi(y)$. Hence $\psi(u) \leqslant \rho(x_n) \leqslant \psi(x_{n+1})$ $+ (1/n) \leqslant \psi(y) + (1/n)$ for all n, so $\psi(u) \leqslant \psi(y)$ – a contradiction.

Proof of Caristi's Theorem Set $\varphi = -\psi$ and for $x, y \in M$ say $x \leqslant y$ provided $\rho(x, y) \leqslant \varphi(x) - \varphi(y)$. Note that by assumption $u < T(u)$ for all $u \in M$. We must verify conditions (a), (b), and (c) of Theorem 2.4. (a) is obvious, and to see that (b) holds observe that if $\{x_n\}$ is any increasing sequence than $\{\varphi(x_n)\}$ is decreasing and bounded below; hence $\{\varphi(x_n)\}$ converges, say to $r \in \mathbb{R}$. This in turn implies that $\{x_n\}$ is a Cauchy sequence. Thus $\{x_n\}$ converges to a point $y \in M$, and since φ is lower semicontinuous it follows that

$$\rho(x_n, y) \leqslant \varphi(x_n) - r \leqslant \varphi(x_n) - \varphi(x).$$

Therefore $x_n \leqslant y$ for all $n \in \mathbb{N}$ proving that (b) holds. Since (c) follows from the fact that φ is bounded below, we conclude that for each $x \in M$ there exists $x' \geqslant x$ such that $T(x') = x'$.

Examples and applications

We begin with the classical Cauchy problem on existence and uniqueness of the solution to a differential equation satisfying a given initial condition.

Example 2.2 Let $f(t, x)$ be a continuous real-valued function defined for t in the interval $[0, T]$, and x in $\mathbb{R}$. The Cauchy initial value problem is the problem of finding a continuously differentiable function x on $[0, T]$ satisfying the differential equation

$$\begin{cases} x'(t) = f(t, x(t)), & t \in [0, T]; \\ x(0) = \xi. \end{cases} \qquad (2.10)$$

The classical result states that if f is lipschitzian with respect to x, i.e., if there exists $L > 0$ such that for all $x, y \in \mathbb{R}$,

$$|f(t, x) - f(t, y)| \leqslant L|x - y|, \qquad t \in [0, T],$$

then the solution to (2.10) exists and is unique. This fact can be proved in many ways. We choose three approaches which serve to illustrate our discussion of the Banach Contraction Principle.

Consider the space $\mathscr{C}[0, T]$ of continuous real-valued functions with standard supremum norm (cf., Example 2.1). Integrating both sides of (2.10) we obtain

$$x(t) = \xi + \int_0^t f(s, x(s)) \, ds.$$

We denote the function defined by the right side of the above by Fx; precisely,

$$(Fx)(t) = \xi + \int_0^t f(s, x(s)) \, ds.$$

Thus $F: \mathscr{C}[0, T] \to \mathscr{C}[0, T]$, and a solution to (2.10) corresponds to a fixed point x of F.

***Approach* A** (This is the approach most commonly presented in textbooks on differential equations.) Observe that for any $x, y \in \mathscr{C}[0, T]$,

$$|(Fx)(t) - (Fy)(t)| = \left| \int_0^t f(s, x(s)) \, ds - \int_0^t f(s, y(s)) \, ds \right|$$

$$\leqslant \int_0^t |f(s, x(s)) - f(s, y(s))| \, ds$$

$$\leqslant \int_0^t L|x(s)) - y(s)| \, ds$$

$$\leqslant Lt \|x - y\|.$$

It follows that

$$\|Fx - Fy\| \leqslant LT \|x - y\|,$$

i.e., $k(F) \leqslant LT$. If $LT < 1$ then the result is immediate via the Banach Contraction Principle. However, if $LT \geqslant 1$, additional steps are needed. Take $h > 0$ such that $Lh < 1$ and consider the space $\mathscr{C}[0, h]$. By replacing T with h in the above argument we obtain a 'local solution' of (2.10), say x_0, in $\mathscr{C}[0, h]$. Now consider the Cauchy problem on $[h, 2h]$:

$$\begin{aligned} x_1'(t) &= f(t, x_1(t)), \\ x_1(h) &= x_0(h). \end{aligned} \tag{2.11}$$

By applying the technique used at the outset to this problem we obtain a unique solution x_1 of (2.11) and, since $x_1(h) = x_0(h)$, x_1 extends x_0 from $[0, h]$ to $[0, 2h]$. This extension is differentiable at h because the Cauchy problem has a *unique* solution in a neighborhood of h. It is now clear that the procedure just described may be repeated on the interval $[2h, 3h]$, and that after a finite number of steps one obtains a solution of (2.10) valid on $[0, T]$.

***Approach* B** (Straightforward evaluations) Repeat the initial calculation and obtain

$$|(F^2x)(t) - (F^2y)(t)| \leqslant \int_0^t |f(s, (Fx)(s)) - f(s, (Fy)(s))| ds$$

$$\leqslant \int_0^t L|(Fx)(s) - (Fy)(s)| ds$$

$$\leqslant \int_0^t L \int_0^s L|x(u) - y(u)| du\, ds$$

$$\leqslant L^2 \int_0^t \int_0^s \|x - y\| du\, ds$$

$$= \frac{L^2 t^2}{2} \|x - y\|.$$

Repetition again yields

$$|(F^3x)(t) - (F^3y)(t)| \leqslant \frac{L^3 t^3}{3!} \|x - y\|,$$

and in general

$$|(F^nx)(t) - (F^ny)(t)| \leqslant \frac{(Lt)^n}{n!} \|x - y\|.$$

Thus $k(F^n) \leqslant (LT)^n/n!$ and the series $\sum_{n=0}^{\infty} k(F^n)$ converges. By our previous observations this implies that F has a unique fixed point x and that for any $x_0 \in \mathscr{C}[0, T]$,

$$\|F^nx_0 - x\| \leqslant \sum_{i=0}^{\infty} \|F^{n+i}x_0 - F^{n+i+1}x_0\|$$

$$\leqslant \sum_{i=0}^{\infty} k(F^{n+i}) \|x_0 - Fx_0\|$$

$$\leqslant R_n \|x - Fx_0\|,$$

where $R_n = \sum_{i=0}^{\infty} (LT)^{n+i}/(n+i)!$ (the nth remainder term in the power series expansion of (e^{LT}).

The advantage of the above approach is that it yields immediate existence of a solution on the whole interval $[0, T]$ along with a good estimate of the rate of convergence of the iterates $\{F^nx_0\}$ to this solution. By applying this procedure to the simple equation $x' = x$, $x(0) = 1$, one sees that this estimate is exact.

Approach C (Remetrization) Since $k(F^n) \leqslant (LT)^n/n!$, $k_\infty(F) = 0$. Thus for any $k > 0$ there exists a metric equivalent to the norm metric on $\mathscr{C}[0, T]$ for which F is a k-contraction. (Such metrics were first introduced by A. Bielecki (1956) and subsequently widely used by others to study a variety of equations.) For a precise illustration, define, for any $\chi \geqslant 0$, a new norm on $\mathscr{C}[0, T]$ as follows:

$$\|x\|_\chi = \max\{[\exp(-\chi Lt)]|x(t)| : t \in [0, T]\}.$$

Then $\|\cdot\|_0$ is the original norm and, since

$$[\exp(-\chi LT)]\|x\|_0 \leqslant \|x\|_\chi \leqslant \|x\|_0, \tag{2.12}$$

all χ-norms are equivalent. Also,

$$\begin{aligned}
|(Fx)(t) - (Fy)(t)| &\leqslant \int_0^t L|x(s) - y(s)|ds \\
&= \int_0^t L \exp(\chi Ls) \exp(-\chi Ls)|x(s) - y(s)|ds \\
&\leqslant \int_0^t L \exp(\chi Ls)\|x - y\|_\chi ds \\
&= \frac{1}{\chi}[\exp(\chi Lt) - 1]\|x - y\|_\chi \\
&\leqslant \frac{1}{\chi}\exp(\chi Lt)\|x - y\|_\chi.
\end{aligned}$$

Multiplying both sides of the above by $\exp(-\chi Lt)$ and taking the maximum on the left side yields

$$\|Fx - Fy\|_\chi \leqslant \frac{1}{\chi}\|x - y\|_\chi.$$

Thus for $\chi > 1$, F is a contraction with respect to the norm $\|\cdot\|_\chi$, and $\lim_{\chi \to \infty} k_{\|\cdot\|_\chi}(F) = 0$.

The above approach can also be used to evaluate the rate of convergence of iterates. If $x, x_0 \in \mathscr{C}[0, T]$ and $x = Fx$, then (for $\chi > 1$) in view of (2.5):

$$\|F^n x_0 - x\|_\chi \leqslant \frac{\chi^{-n}}{1 - \chi^{-1}}\|x_0 - Fx_0\|_\chi,$$

and using (2.12),

$$\|x - F^n x_0\| \leqslant \exp(\chi LT)\frac{\chi^{-n}}{1 - \chi^{-1}}\|x_0 - Fx_0\|. \tag{2.13}$$

Choosing $\chi = n/LT$ yields a nice evaluation of the rate of convergence in (2.13):

$$\|x - F^n x_0\| \leqslant \left[\frac{eLT}{n}\right]^n \frac{n}{n - LT} \|x_0 - Fx_0\|.$$

Example 2.3 A basic fact in theory of functions of a complex variable is that 'holomorphic mappings do not increase hyperbolic distances'. To understand this, let D be the open unit disc in the complex plain $\mathbb{C}$:

$$D = \{z \in \mathbb{C} : |z| < 1\}.$$

The so-called *hyperbolic* (or Poincaré) *metric* ρ on D is defined as follows: For $z, w \in D$, set

$$\rho(z, w) = \tanh^{-1}\left|\frac{z - w}{1 - z\bar{w}}\right|.$$

The space (D, ρ) is unbounded and complete, and it serves as a model for Lobachewski's (hyperbolic) geometry. Arcs of circles orthogonal to the unit circle ∂D are isometric to the real line $\mathbb{R}$ and form the system of geodesics (hyperbolic lines) in this model. Let $f : D \to D$ be holomorphic (thus representable by a power series). The quoted statement above means that $k(f) \leqslant 1$, i.e.,

$$\rho(f(z), f(w)) \leqslant \rho(z, w), \qquad z, w \in D.$$

(There is also a nice class of holomorphic ρ-isometries (automorphisms) on D consisting of all mappings h of the form:

$$h(z) = \exp(i\theta)\frac{z + a}{1 + z\bar{a}},$$

where $a \in D$ is fixed.)

Any holomorphic mapping which maps D 'strictly inside' D in the sense

$$\sup\{|f(z)| : z \in D\} < 1$$

is a ρ-contraction, a fact which implies the following:

Theorem 2.5 *Any holomorphic mapping* $f : D \to D$ *such that* $\sup\{|f(z)| : z \in D\}$ < 1 *has a unique fixed point in* D.

The above theorem can be easily extended to more general settings. By the Riemann Mapping Theorem, any simply connected domain U in the complex plane whose boundary consists of more than one point is conformally equivalent to D. This means that there exists a holomorphic univalent (one-to-one) mapping $g: D \to U$ with $g(D) = U$. Any such mapping generates a metric ρ_U on U:

$$\rho_U(z, w) = \rho(g^{-1}(z), g^{-1}(w)).$$

The space (U, ρ_U) is isometric to (D, ρ) and g is an isometry. Also, observe that any holomorphic mapping $f: U \to U$ generates a holomorphic mapping $\tilde{f}: D \to D$ via the formula:

$$\tilde{f}(x) = g^{-1} \circ f \circ g(z).$$

If z is a fixed point of $\tilde{f}$ then clearly $g(z)$ is a fixed point of f. The condition

$$\sup\{|\tilde{f}(z)|: z \in D\} < 1$$

becomes: $f(U)$ is ρ_U-bounded in (U, ρ_U). Thus we have the following:

Theorem 2.6 *Any holomorphic mapping $f: U \to U$ such that $f(U)$ is bounded in the metric ρ_U has a unique fixed point.*

Remark If $f(U)$ is ρ_U-bounded then $f(U)$ lies strictly inside U in the sense that $f(U)$ is separated from the boundary ∂U, of U (with the convention that ∂U contains the 'point at infinity' for unbounded U).

The approach sketched above has been widely developed and extended not only to multidimensional settings, but also to Banach space settings. (See, e.g., Goebel and Reich, 1984).

Example 2.4 (Self-similar sets) Let (M, ρ) be a complete metric space, let $\mathcal{M}$ denote the family of all nonempty, bounded, closed subsets of M, and let $\mathcal{N}$ denote the subfamily of $\mathcal{M}$ consisting of all compact sets. For $X, Y \in \mathcal{M}$, set

$$d(X, Y) = \sup\{\text{dist}(y, X): y \in Y\};$$
$$d(Y, X) = \sup\{\text{dist}(x, Y): x \in X\};$$

and let

$$D(X, Y) = \max\{d(X, Y), d(Y, X)\}.$$

D provides a metric for $\mathcal{M}$ (hence $\mathcal{N}$) commonly called the *Hausdorff metric*. It is an elementary, yet rather technical, exercise to verify that completeness of M implies completeness of $(\mathcal{M}, D)$ and $(\mathcal{N}, D)$ (see, e.g., Blumenthal, 1953).

Now suppose $T_1, T_2, \ldots, T_n$ is a finite family of ρ-contractions on M.

These mappings generate a mapping $\mathcal{F}: \mathcal{N} \to \mathcal{N}$ by taking

$$\mathcal{F}(X) = \bigcup_{i=1}^{n} T_i(X), \qquad X \in \mathcal{N},$$

and it is not difficult to see that $\mathcal{F}$ is a contraction on $\mathcal{N}$ relative to D with $k(\mathcal{F}) \leqslant \max\{k(T_i): i = 1, \ldots, n\}$. Hence we have the following.

Theorem 2.7 *Let M be a complete metric space and let $T_i: M \to M$, $i = 1, \ldots, n$, be a family of contractions. Then there is a unique, nonempty, compact subset X of M such that*

$$X = \bigcup_{i=1}^{n} T_i(X). \tag{2.14}$$

This theorem has applications in the interesting modern theory of 'fractals' (see, e.g., Peitgen and Richter, 1986). In particular, if M is a euclidean space $\mathbb{R}^p$ and the mappings T_i geometric similarities with respective scales $k_i < 1$, then the set satisfying (2.14) is called *self-similar* with respect to $\{T_1, \ldots, T_n\}$. Sometimes such sets are 'very exotic' (and called fractals).

For an interesting special case of the above, consider the real line $\mathbb{R}$ and two similarities defined as follows:

$$T_1 x = \tfrac{1}{3} x; \qquad T_2 x = \tfrac{1}{3} x + \tfrac{2}{3}.$$

The mapping $\mathcal{F}$ is defined by associating with each compact $X \subset \mathbb{R}$ the set

$$\mathcal{F}(X) = \tfrac{1}{3} X \cup (\tfrac{1}{3} X + \tfrac{2}{3}).$$

Since $\mathcal{F}$ is a contraction relative to the Hausdorff metric D we may obtain its fixed point by iteration. Take $X_0 = [0, 1]$; then $X_1 = [0, \tfrac{1}{3}] \cup [\tfrac{2}{3}, 1]$; $X_2 = [0, \tfrac{1}{9}] \cup [\tfrac{2}{9}, \tfrac{3}{9}] \cup [\tfrac{6}{9}, \tfrac{7}{9}] \cup [\tfrac{8}{9}, 1]$, etc. The sequence $\{X_n\}$ converges in D to the well-known Cantor set C.

Example 2.5 (Square roots in Banach algebras) Recall that a Banach algebra X is a Banach space $(X, \| \ \|)$ in conjunction with a product operator which satisfies for $x, y, z \in X, \alpha \in \mathbb{R}$,

$$x(yz) = (xy)z; \qquad x(y+z) = xy + xz; \qquad (y+z)x = yx + zx;$$
$$(\alpha x)y = \alpha(xy) = x(\alpha y);$$

in addition to the norm inequality

$$\|xy\| \leqslant \|x\| \cdot \|y\|.$$

For any $z \in X$, let $X(z)$ denote the subalgebra generated by z (the smallest closed subalgebra of X which contains z). The algebra $X(z)$ is always commutative: $xy = yx$ for all $x, y \in X(z)$.

We assert that for any $z \in X$ with $\|z\| < 1$ there exists a unique element $x \in X(z)$ such that $\|x\| < 1$ and

$$x^2 - 2x + z = 0. \tag{2.15}$$

To see this, take any number satisfying $\|z\| < d < 1$ and consider the mapping T defined by

$$Tx = \tfrac{1}{2}(x^2 + z), \qquad x \in B(0; d) \subset X(z).$$

Since $\|Tx\| \leqslant \tfrac{1}{2}(\|x\|^2 + \|z\|) \leqslant \tfrac{1}{2}(d^2 + d) < d$, $T: B(0; d) \to B(0; d)$. Moreover, since $X(z)$ is commutative,

$$\begin{aligned}
\|Tx - Ty\| = \tfrac{1}{2}\|x^2 - y^2\| &= \tfrac{1}{2}\|(x - y)(x + y)\| \\
&\leqslant \tfrac{1}{2}(\|x\| + \|y\|)\|x - y\| \\
&\leqslant d\|x - y\|.
\end{aligned}$$

This proves that T is a contraction mapping on $B(0; d)$ and hence has a *unique* fixed point x in $B(0; d)$. Since $d < 1$ can be chosen arbitrarily near 1, x is the unique fixed point of T, hence the unique solution of (2.15), in the interior of $B(0; 1)$.

If X has a unit e which satisfies $ex = xe = x$ for all $x \in X$, then (2.15) can be written in the form

$$(e - x)^2 = e - z$$

and the assertion may be reformulated as follows: For any element of X of the form $e - z$ with $\|z\| < 1$ there exists a unique element $y = e - x$ with $x \in X(z)$ and $\|x\| < 1$ such that $y^2 = e - z$. In other words $e - z$ has a 'square root'.

We remark that the above fact may be proved under the less restrictive assumption that the spectral radius $r(z)$ of z is less than 1.

Since

$$r(z) = \lim_{n \to \infty} \|z^n\|^{1/n} = \inf_n \|z^n\|^{1/n},$$

$r(z)$ is a counterpart of the constant $k_\infty(T)$ introduced earlier. If X is the finite dimensional Banach algebra of the $n \times n$ complex matrices, it is known that the spectral radius $r(A)$ of the matrix A is the number $\max\{|\lambda_i|: i = 1, \ldots, n\}$ where $\{\lambda_1, \ldots, \lambda_n\}$ is the set of all eigenvalues of A. Thus the assertion of Example 2.5 specializes to: a matrix of the form $I - A$ where each eigenvalue λ of A satisfies $|\lambda| < 1$ has a square root of the same form.

Another application of the above is the following. Let S be a set and let X denote an algebra of bounded real-valued functions defined on S equipped with the usual supremum norm. (The product operation on X is pointwise multiplication.) Then it can be shown that if X is complete and contains the unit function, any nonnegative function $f \in X$ has a square root $f^{1/2} \in X$. (Verification that this is a consequence of the observations of Example 2.5 is left to the reader.) Consequently for $f, g \in X$ the following all belong to X:

$$|f| = (f^2)^{1/2};$$

$$\max\{f, g\} = \tfrac{1}{2}(f + g + |f - g|);$$

$$\min\{f, g\} = \tfrac{1}{2}(f + g - |f - g|).$$

Observe that the above facts are valid without assuming that f has any special properties, or even that S has a topology. Nevertheless they are crucial to the proof of the Stone–Weierstrass Theorem, thus showing that the Contraction Mapping Principle is indirectly involved in establishing one of the most useful theorems in all of analysis. For the sake of completeness, and since we shall invoke this theorem later, we include the details.

Theorem 2.8 (Stone–Weierstrass Theorem) *Let S be a compact Hausdorff topological space and let $\mathscr{C}(S)$ be the algebra of all bounded, continuous, real-valued functions on S. Suppose X is a subalgebra of $\mathscr{C}(S)$ which contains the unit function and which separates points of S. Then X is dense in $\mathscr{C}(S)$.*

Proof Recall first that since X separates points of S, for any $x, y \in S$ with $x \neq y$ there exists $f \in X$ such that $f(x) \neq f(y)$.

Now let $f \in \mathscr{C}(S)$, select $\epsilon > 0$, and fix $x \in S$. Since X separates points of S and contains the constant functions, for each $y \in S$ there exists a function $g_y \in X$ such that $g_y(x) = f(x)$ and $g_y(y) = f(y)$. To see this, select $g \in S$ such that $g(x) \neq g(y)$ and for $u \in S$, set

$$g_y(u) = f(x) + (f(y) - f(x))(g(y) - g(x))^{-1}(g(u) - g(x)).$$

Since g_y is continuous, there exists a neighborhood U_y of y such that if $t \in U_y$, $g_y(t) < f(t) + \epsilon$. The neighborhoods $\{U_y\}$ cover S so, since S is compact, there exists $\{y_1, \ldots, y_n\}$ such that $S \subset \bigcup_{i=1}^n U_{y_i}$. Set

$$h_x(t) = \min\{g_{y_1}(t), \ldots, g_{y_n}(t)\}.$$

Then $h_x(t) < f(t) + \epsilon$ for all $t \in S$, and $h_x \in \bar{X}$ where $\bar{X}$ denotes the smallest complete subalgebra of $\mathscr{C}(S)$ which contains X. (Recall, X is not assumed complete.)

Now for each $x \in S$ choose a neighborhood V_x of x such that $h_x(t) > f(t) - \epsilon$ for all $t \in V_x$, and let $\{V_{x_1}, \ldots, V_{x_m}\}$ be a finite subcovering of the covering $\{V_x\}$ of S. Set

$$h(t) = \max\{h_{x_1}(t), \ldots, h_{x_m}(t)\},$$

and observe that $h \in \bar{X}$ with $\|f - h\| < \epsilon$. Since $\epsilon > 0$ was arbitrary, the proof is complete.

Example 2.6 Banach's Contraction Principle can also be used to establish the following important fact.

Theorem 2.9 *Let U be an open subset of a Banach space X and let $T: U \to X$ be a contraction mapping. Then if $F = I - T$, the set $F(U)$ is open.*

Proof Let $z \in U$ and select $r > 0$ so that $B(z; r) \subset U$. Now select $y \in X$ so that $\|F(z) - y\| < (1 - k)r$ where $k = k(T) < 1$. We complete the proof by showing that y is in the range of $B(z; r)$ under F. Let the mapping f be defined by $f(x) = x - (F(x) - y)$, $x \in B(z; r)$. Then:

$$\begin{aligned}
\|f(x) - z\| &= \|T(x) + y - z\| \\
&\leqslant \|T(x) - T(z)\| + \|T(z) + y - z\| \\
&\leqslant k\|x - z\| + \|F(z) - y\| \\
&\leqslant kr + (1 - k)r = r.
\end{aligned}$$

Thus $f: B(z; r) \to B(z; r)$, and since $k(f) = k(T) = k < 1$ it follows from Banach's Principle that f has a fixed point $w \in B(z; r)$ from which $y = F(w)$, completing the proof.

We finish with an application of Caristi's theorem.

Example 2.7 In 1928 K. Menger introduced the following concept of metric convexity.

Definition A metric space (M, ρ) is said to be *metrically convex* if for any two points $x, y \in M$ with $x \neq y$ there exists $z \in M$, $x \neq z \neq y$, such that

$$\rho(x, z) + \rho(z, y) = \rho(x, y).$$

If z is as above we say that z *lies between* x and y and denote this fact by the symbol (xzy). This is a concept fundamental to axiomatic metric geometry (e.g., see Blumenthal (1953)).

The fundamental theorem of metric convexity is the following.

Theorem (*Menger*) *If (M, ρ) is a complete and metrically convex metric space, then any two points $x, y \in M$ are joined by a metric segment, i.e., there exists an isometry $\varphi: [0, \rho(x, y)] \to M$ with $\varphi(0) = x$ and $\varphi(\rho(x, y)) = y$.*

Our proof is basically the same as the original (Menger, 1928) except for the proof of the following lemma, which utilizes Caristi's theorem instead of a lengthy transfinite induction. (We should remark that a different proof of Menger's theorem, due to Aronszajn may be found, for example, in Blumenthal (1973).)

Lemma 2.1 *Let (M, ρ) be a complete metric space with $x, y \in M$, $x \ne y$, and suppose $0 < \lambda < \rho(x, y)$. Let*

$$B(x, y) = \{z \in M : (xzy)\},$$
$$S = S(x, y, \lambda) = \{z \in B(x, y) : \rho(x, z) \le \lambda\} \cup \{x\}.$$

Then there exists a point $z_\lambda \in M$ such that

(a) $z_\lambda \in S(x, y, \lambda)$,
(b) $u \in B(x, y)$ *and* $(xz_\lambda u)$ *implies* $\rho(x, u) > \lambda$.

Proof

Case 1. There exists $z' \in S$ with $\rho(x, z') < \lambda$ such that $(xz'u) \Rightarrow u \notin S$. In this case take $z_\lambda = z'$.

Case 2. For each $z \in S$ with $\rho(x, z) < \lambda$ there exists y_z such that (xzy_z). In this case define $G: S \to S$ by taking $G(z) = y_z$ if $\rho(x, z) < \lambda$ and $G(z) = z$ otherwise. Define $\varphi: S \to \mathbb{R}^+$ by taking $\varphi(z) = \lambda - \rho(x, z)$. Then clearly φ is continuous and for $z \in S$,

$$\rho(z, G(z)) = \rho(x, G(z)) - \rho(x, z) = \lambda - \rho(x, z) - (\lambda - \rho(x, G(z))) = \varphi(z) - \varphi(G(z)).$$

Since S is closed (hence complete), by Caristi's theorem $G(z') = z'$ for some point $z' \in S$. This implies $\rho(x, z') = \lambda$ and so $z' = z_\lambda$ satisfies (a) and (b).

We shall need the following transitivity relationship for betweenness in metric spaces (see Blumenthal, 1953, p. 33) which asserts that for points $p, q, r, s \in M$,

$$((pqr) \quad \text{and} \quad (prs)) \Leftrightarrow ((pqs) \quad \text{and} \quad (qrs)).$$

Lemma 2.2 *Let M be complete and convex, $x, y \in M, x \neq y$, and suppose $0 < \lambda < \rho(x, y)$. Then there exists $z' \in M$ such that $(xz'y)$ and $\rho(x, z') = \lambda$.*

Proof By Lemma 2.1 there exists $z_\lambda \in M$ such that:

(a) $z_\lambda \in S(x, y, \lambda)$,
(b) $u \in B(x, y)$ and $(xz_\lambda u)$ imply $\rho(x, u) > \lambda$.

Let $\lambda' = \rho(x, y) - \lambda$ and again apply Lemma 2.1 to obtain $y_{\lambda'} \in M$ such that:

(a)' $y_{\lambda'} \in S(y, z_\lambda, \lambda')$,
(b)' $u \in B(y, z_\lambda)$ and $(yy_{\lambda'} u)$ imply $\rho(y, u) > \lambda'$.

Case 1. $z_\lambda = y_{\lambda'}$. Then, since $\rho(x, y) = \rho(x, z_\lambda) + \rho(z_\lambda, y)$, it follows that $\rho(x, z_\lambda) = \lambda$.

Case 2. $z_\lambda \neq y_{\lambda'}$. In this case, since M is convex, there exists $w \in M$ such that: $(z_\lambda w y_{\lambda'})$. By assumption the relations $(xz_\lambda y), (z_\lambda y_{\lambda'} y)$, and $(z_\lambda w y_{\lambda'})$ hold. It follows from transitivity of betweenness that $(xwy), (xz_\lambda w), (ywz_\lambda)$, and $(yy_{\lambda'} w)$ also hold. Now, (xwy) and $(xz_\lambda w)$ imply $\rho(x, w) > \lambda$ by (b), while (ywz_λ) and $(yy_{\lambda'} w)$ imply $\rho(y, w) > \lambda'$ by (b)'. Therefore $\rho(x, y) = \rho(x, w) + \rho(w, y) > \lambda + \lambda' = \rho(x, y)$. This contradiction establishes Lemma 2.2 via case 1.

Proof of Menger's theorem Let $x_0, x_1 \in M, x_0 \neq x_1$. By Lemma 2.2 there exists $x_{1/2} \in M$ such that $\rho(x_0, x_{1/2}) = \rho(x_{1/2}, x_1) = \frac{1}{2}\rho(x_0, x_1)$ (i.e., $x_{1/2}$ is a 'midpoint' of the pair (x_0, x_1)). Let $d = \rho(x_0, x_1)$ and define the mapping F by taking

$$F(0) = x_0, F(d/2) = x_{1/2}, F(d) = x_1.$$

Again by Lemma 2.2 there exist points $x_{1/4}, x_{3/4}$ which are respective midpoints of the pairs $(x_0, x_{1/2})$ and $(x_{1/2}, x_1)$. Set

$$F(d/4) = x_{1/4}, F(3d/4) = x_{3/4},$$

and use transitivity of betweenness to conclude that F is an isometry on the set $\{0, d/4, d/2, 3d/4, d\}$. By induction it is possible to obtain points $\{x_{p/2^n}\}$, $1 \leqslant p \leqslant 2^n - 1$ $(n = 1, 2 \ldots)$, in M such that the mapping $F: pd/2^n \to x_{d/2^n}$ is an isometry. Since $\{pd/2^n\}$ is a dense subset of $[0, d]$ and since M is complete, it is possible to extend F in the obvious way to the entire interval $[0, d]$ and thus obtain a metric segment in M joining x_0 and x_1. This completes the proof.

3

Nonexpansive mappings: introduction

As before, let M denote a metric space (with metric ρ) and suppose $D \subset M$. A mapping $T: D \rightarrow M$ is called *nonexpansive* if its Lipschitz constant $k(T)$ does not exceed 1. Thus this class of mappings includes the contractions and strictly contractive mappings; moreover it contains all isometries (including the identity). Explicitly, $T: D \rightarrow M$ is nonexpansive if

$$\rho(Tx, Ty) \leqslant \rho(x, y), \qquad x, y \in M. \tag{3.1}$$

We have already noted that nonexpansive mappings may be fixed point free, and obviously when such a mapping has a fixed point it need not be unique (e.g., the identity mapping). The theory for nonexpansive mappings is fundamentally different from that of the contraction mappings in another sense. Even if a nonexpansive mapping has a fixed point the iterative procedure described in the previous section generally fails to converge.

The following are among the most basic questions asked in the study of nonexpansive mappings. We shall eventually provide at least partial answers to each.

(a) What additional assumptions must be added regarding the structure of the space M and/or restrictions on T to assure the existence of at least one fixed point?

(b) What is the structure of the set of all fixed points of a given T?

(c) What can be said about the behavior of the iterates of T?

Very little of interest can be said about the fixed point theory of nonexpansive mappings within the general metric space framework. Because of this we shall primarily confine our attention to a Banach space setting (certainly a setting sufficiently general to include interesting applications).

Let X be a Banach space with norm $\|\cdot\|$ and let K denote a nonempty, closed convex and bounded subset of X. In this context a mapping $T: K \rightarrow K$ is nonexpansive if

$$\|Tx - Ty\| \leqslant \|x - y\|, \qquad x, y \in K. \tag{3.2}$$

Note that (3.2) clearly implies that all the iterates T^n of T, $n = 0, 1, 2, \ldots$ (with $T^0 = I$) are nonexpansive as well.

We begin by noting a basic elementary fact:

Lemma 3.1 *Under the above assumptions on T and K*, $\inf\{\|x - Tx\| : x \in K\} = 0$.

Proof Fix $z \in K$ and $\epsilon \in (0, 1)$, and consider the mapping $T_\epsilon : K \to K$ defined by

$$T_\epsilon x = \epsilon z + (1 - \epsilon)Tx.$$

T_ϵ is a contraction since

$$\|T_\epsilon x - T_\epsilon y\| \leqslant (1 - \epsilon)\|Tx - Ty\| \leqslant (1 - \epsilon)\|x - y\|, \qquad x, y \in K,$$

and in view of this there exists $x_\epsilon \in K$ for which $x_\epsilon = T_\epsilon x_\epsilon$. Hence

$$\begin{aligned}\|x_\epsilon - Tx_\epsilon\| &= \|\epsilon z + (1 - \epsilon)Tx_\epsilon - Tx_\epsilon\| \\ &= \epsilon\|z - Tx_\epsilon\| \\ &\leqslant \epsilon \text{ diam } K.\end{aligned}$$

The result follows upon letting $\epsilon \to 0$.

Lemma 3.1 asserts that T has 'almost fixed points' in the sense that points may be found (constructively) which are moved an arbitrarily small distance. Consequently there always exist sequences $\{y_n\}$ in K for which $\lim_{n \to \infty} \|y_n - Ty_n\| = 0$. We shall call such sequences 'approximate fixed point sequences',

In view of the above, the question of whether T has a fixed point is equivalent to the question of whether the continuous function $\varphi : K \to \mathbb{R}$ defined by $\varphi(x) = \|x - Tx\|$ attains its minimal value zero. The answer is yes if K is compact.

Theorem 3.1 *If K is a nonempty, compact and convex subset of a Banach space, then any nonexpansive mapping of K into K has a fixed point.*

Of course the above is a special case of the fundamental Schauder Fixed Point Theorem which asserts that any *continuous* $f : K \to K$ has a fixed point. However, in contrast to the above, Schauder's Theorem has 'nonmetric' features (cf. Chapter 18).

As noted in Example 2.1, the above theorem is not true in general for noncompact K.

Before discussing additional examples we recall that a subset D of K is invariant under $T : K \to K$ if $T(D) \subset D$. The T-invariant subsets of K which are nonempty, closed and convex are of particular interest since the search for fixed points of T may be confined to such sets. Also, the intersection of any family of nonempty, closed, convex T-invariant subsets of K is itself closed, convex and T-invariant, although it may be empty.

For a given $T: K \to K$ a descending sequence of nonempty, closed, convex T-invariant subsets may be obtained by setting

$$K_0 = K; \qquad K_{n+1} = \overline{\mathrm{conv}}\, T(K_n), \qquad n = 0, 1, \dots.$$

The set $K_\infty = \bigcap_{n=0}^\infty K_n$ is closed, convex and T-invariant. However it may be the case that $K_\infty = \emptyset$, or that the sequence stabilizes: $K_p = K_{p+1} = \dots$.

We now turn to some examples.

Examples 3.1 (Cf., Example 2.1 of Chapter 2) Let $X = \mathscr{C}[0, 1]$ and set

$$K = \{x \in X : 0 = x(0) \leqslant x(t) \leqslant x(1) = 1\}.$$

Define $T: K \to K$ by $(Tx)(t) = tx(t)$. As previously noted, T is contractive and fixed point free. If $\{K_n\}$ is defined as above ($K_{n+1} = \overline{\mathrm{conv}}\, T(K_n), n = 0, 1, \dots$) then for each n, $K_n \subset \{x \in K : x(t) \leqslant t^n\}$ and consequently, $\bigcap_{n=1}^\infty K_n = \emptyset$. Also, since the iterates of T at $x \in K$ are given by $(T^n x)(t) = t^n x(t)$, we have the 'strange' fact: For any $x, y \in K$,

$$\lim_{n \to \infty} \| y - T^n x \| = 1 = \mathrm{diam}\, K.$$

Example 3.2 Let $X = l^1 (= l^1(\mathbb{N}))$ and let $\{e^n\} = \{\delta_{in}\}$ be the standard basis for l^1 (where e^n stands for the vector with 1 in the nth position, zeros elsewhere). Consider the set

$$K = \mathrm{conv}\{e^n : n \geqslant 1, 2, \dots\} = \{x = \{x_i\} : x_i \geqslant 0, i = 1, 2, \dots, ; \|x\| = 1\}.$$

Then diam $K = 2$ and the shift operator S defined by

$$Sx = S(x_1, x_2, \dots) = (0, x_1, x_2, \dots)$$

is an isometry mapping K into K which is fixed point free. Indeed, if $Sx = x$, $x = 0$ contradicting $\|x\| = \Sigma_{i=1}^\infty x_i = 1$. Also, as before, the sets $K_{n+1} = \overline{\mathrm{conv}}\, S(K_n), n = 1, 2, \dots$, form a decreasing sequence with empty intersection, while for any $x, y \in K$, $\lim_{n \to \infty} \| y - S^n x \| = 2 = \mathrm{diam}\, K$.

On the other hand, the analogous set K in $l^1(\mathbb{Z})$,

$$K = \left\{ x = \{x_n\}_{n=-\infty}^\infty \in l^1(\mathbb{Z}) : x_n \geqslant 0, n = 0, \pm 1, \dots; \sum_{n=-\infty}^{+\infty} x_n = 1 \right\},$$

behaves differently under the isometry (shift) $S : x = \{x_n\} \to Sx = \{x_{n-1}\}$. It is still the case that for any $x, y \in K$, $\lim_{n \to \infty} \| S^n x - y \| = 2 = \mathrm{diam}\, K$, but since $S(K) = K$, also $S^n(K) = K$.

Examples similar to the above may be obtained in such standard spaces as $c_0(\mathbb{N}), c_0(\mathbb{Z}), L^1(0, \infty), L^1(-\infty, +\infty)$, or in spaces of continuous functions vanishing at infinity.

We remark that in the specific examples discussed above, the sets K are subsets of the unit sphere in X, and they possess a type of irregular behavior; namely, for each $x \in K$,

$$\sup\{\|x - y\| : y \in K\} = \operatorname{diam} K.$$

Similar examples exist in the unit ball:

Example 3.3 In the space $c_0(\mathbb{N})$ the isometry T defined by

$$T(x_1, x_2, \ldots) = (1, x_1, x_2, \ldots)$$

maps the unit ball into its boundary without having a fixed point.

Example 3.4 For $X = \mathscr{C}[-1, 1]$ define the mapping T by

$$(Tx)(t) = \min\{1, \max\{-1, x(t) + 2t\}\}.$$

It is trivial to see that T is nonexpansive and maps the unit ball into its boundary. Moreover, since either $(Tx)(t) > x(t)$ for some $t > 0$ or $(Tx)(t) < x(t)$ for some $t < 0$, T cannot have a fixed point.

While the above examples might suggest that generally for noncompact, convex sets K nonexpansive mappings $T: K \to K$ exist which do not have fixed points, we shall see that this is far from true. We begin with a specific observation about l^1 $(= l^1(\mathbb{N}))$.

A bounded sequence $\{x^n\}$ of points of l^1 is coordinate-wise convergent to $x \in l^1$ if for any $i = 1, 2, \ldots$, $\lim_{n \to \infty} x_i^n = x_i$. For a fixed sequence $\{x^n\}$ in l^1 and arbitrary $y \in l^1$, denote

$$r(y) = \limsup_{n \to \infty} \|x^n - y\|.$$

Moreover, let P_i denote the standard projections onto the subspaces spanned by $\{e^1, e^2, \ldots, e^i\}, i = 1, 2, \ldots$, and for each i let $Q_i = I - P_i$. Thus

$$P_i(x_1, x_2, \ldots) = (x_1, x_2, \ldots, x_i, 0, 0, \ldots);$$
$$Q_i(x_1, x_2, \ldots) = (0, 0, \ldots, 0, x_{i+1}, x_{i+2}, \ldots).$$

Lemma 3.2 *If $\{x^n\}$ is a bounded sequence in l^1 converging coordinate-wise to x, then for any $y \in l^1$,*

$$r(y) = r(x) + \|x - y\|. \tag{3.3}$$

Proof Notice first that for any $i = 1, 2, \ldots,$

$$r(x) = \limsup_{n \to \infty} \|Q_i(x^n - x)\|;$$

$$\|x - y\| = \lim_{i \to \infty} \lim_{n \to \infty} \|P_i(x^n - y)\|;$$

$$\lim_{i \to \infty} \|Q_i(x - y)\| = 0.$$

The result then follows from the inequalities below upon passing to the limit first with n, then with i:

$$\|P_i(x^n - y)\| + \|Q_i(x^n - x)\| - \|Q_i(x - y)\| \leqslant \|P_i(x^n - y)\| + \|Q_i(x^n - y)\|$$
$$= \|x^n - y\|$$
$$\leqslant \|P_i(x^n - y)\| + \|Q_i(x^n - x)\|$$
$$+ \|Q_i(x - y)\|.$$

Example 3.5 Consider the set $K_\epsilon \subset l^1$ defined for $\epsilon \in (0, 1)$ by:

$$K_\epsilon = \overline{\mathrm{conv}}\{(1 - \epsilon)e^1, e^2, e^3, \ldots\}.$$

K_ϵ consists of all points of the form $((1 - \epsilon)\alpha_1, \alpha_2, \alpha_3, \ldots)$ where each $\alpha_i \geqslant 0$ and $\Sigma \alpha_i = 1$; equivalently, $x = (x_1, x_2, \ldots) \in K_\epsilon$ if $x_i \geqslant 0, i = 1, 2, \ldots,$ and $(1 - \epsilon)^{-1} x_1 + \Sigma_{i=2}^{\infty} x_i = 1$.

Suppose $T: K_\epsilon \to K_\epsilon$ is nonexpansive. As in Lemma 3.1, there exists a sequence $\{x^n\}$ in K_ϵ for which $\lim_{n \to \infty} \|x^n - Tx^n\| = 0$. Since any bounded sequence in l^1 contains a coordinate-wise convergent subsequence we may assume $\{x^n\}$ is itself coordinate-wise convergent with limit, say x. While x need not belong to K_ϵ it does lie in the convex envelope of K_ϵ and $\{0\}$ ($\mathrm{conv}(K_\epsilon \cup \{0\})$). This means x is of the form

$$x = ((1 - \epsilon)\beta_1, \beta_2, \beta_3, \ldots)$$

with $\beta_1 \geqslant 0$ and $\Sigma_{i=1}^{\infty} \beta_i \leqslant 1$. If $x \in K_\epsilon$, i.e., if $\Sigma_{i=1}^{\infty} \beta_i = 1$ then, since

$$r(Tx) = r(x) + \|Tx - x\|$$

and

$$r(Tx) = \limsup_{n \to \infty} \|Tx - x^n\|$$

$$\leqslant \limsup_{n \to \infty} \|Tx - Tx^n\| + \limsup_{n \to \infty} \|Tx^n - x^n\|$$

$$\leqslant \limsup_{n \to \infty} \|x - x^n\|$$

$$= r(x),$$

we conclude $x = Tx$. On the other hand, if $x \notin K_\epsilon$, i.e., if $1 - \Sigma_{i=1}^{\infty} \beta_i = \delta > 0$, then the (unique) point of K_ϵ which is nearest x is

$$z = ((1-\epsilon)(\beta_1 + \delta), \beta_2, \beta_3, \dots).$$

Thus

$$(1-\epsilon)\delta = \|z - x\| = \inf\{\|x - y\| : y \in K_\epsilon\}$$

and in view of (3.3), for any $y \in K_\epsilon$,

$$r(y) = r(x) + \|y - x\| \geqslant r(x) + \|z - x\| = r(z).$$

But, as before,

$$r(Tz) \leqslant r(z).$$

Since the function r attains its minimum only at z, $Tz = z$. *Thus any nonexpansive mapping $T: K_\epsilon \to K_\epsilon$ has a fixed point.* Notice that for $\epsilon > 0$ small, K_ϵ differs only slightly from K_0; in the Hausdorff metric, $D(K_\epsilon, K_0) = \epsilon$. Since there are nonexpansive mappings of K_0 into K_0 without fixed points (the shift operator of Example 3.3), the present example illustrates a lack of 'stability' among sets having the fixed point property for nonexpansive self-mappings.

In connection with the above we observe that the modified shift S_ϵ defined by

$$S_\epsilon(x_1, x_2, \dots) = (0, (1-\epsilon)^{-1} x_1, x_2, \dots)$$

maps K_ϵ into itself, but this mapping is not nonexpansive. However, if we replace the classical norm with the equivalent norm $\|\cdot\|_\epsilon$, where

$$\|x\|_\epsilon = (1-\epsilon)^{-1} |x_1| + \sum_{i=2}^{\infty} |x_i|,$$

then S_ϵ becomes an $\|\cdot\|_\epsilon$-isometry and plays the role of a standard fixed point free shift in K_ϵ. Dual to this observation is the fact that with a slight change in the original norm we can obtain a new norm relative to which all nonexpansive self-mappings of K_0 have fixed points.

The preceding examples suggest that a nonexpansive mapping T may frequently fail to have a fixed point in a setting which permits the existence of a decreasing sequence of nonempty, closed, convex, T-invariant sets having empty intersection. However, since closed, convex sets are also closed in the weak topology, this situation cannot occur in a weakly compact setting. We turn briefly to this setting now and continue in greater depth in the next chapter.

Definition 3.1 A nonempty, closed, convex subset D of a given set K is said to be a *minimal invariant set* for a mapping $T: K \to K$ if $T(D) \subset D$ and if D has no nonempty, closed and convex proper subsets which are T-invariant.

We shall subsequently modify the above definition in obvious ways so that it applies to narrower classes of sets (e.g., minimal T-invariant weakly (or weak*) compact sets).

The following theorem is basic to much of what follows:

Theorem 3.2 *Suppose K is a nonempty, weakly compact, convex subset of a Banach space. Then for any mapping $T: K \to K$ there exists a closed convex subset of K which is minimal T-invariant.*

Proof Consider the family $\mathcal{M}$ of all nonempty, closed and convex (thus weakly compact) subsets of K which are T-invariant, and order this family by set inclusion: For $K_1, K_2 \in \mathcal{M}$, $K_1 \leqslant K_2$ provided $K_2 \subset K_1$. By weak compactness, any chain (linearly ordered family) of sets in $\mathcal{M}$ has nonempty intersection, hence an upper bound relative to $\leqslant$. By Zorn's Lemma there exists at least one set $D \in \mathcal{M}$ which is maximal relative to $\leqslant$, hence minimal T-invariant.

The above theorem has an obvious analogue for weak* closed, convex subsets of a weak* compact set in a dual Banach space. In such settings, only minimal T-invariant sets need to be examined in seeking the existence of fixed points of T. A useful basic property of such sets is given by the following:

Lemma 3.3 *If K is a nonempty, closed, convex and minimal T-invariant set, then*

$$K = \overline{\text{conv}} \ T(K).$$

Proof Obviously $\overline{\text{conv}} \ T(K)$ is closed, convex and T-invariant. By minimality, it cannot be a proper subset of K.

If a mapping $T: K \to K$ has a fixed point x, then clearly $\{x\}$ is closed, convex and minimal T-invariant. A question which remained open for many years in the theory of nonexpansive $T: K \to K$ was whether, for weakly compact, convex K, such a set could exist consisting of more than one point. D. E. Alspach settled this question affirmatively in 1981 by producing just such a set in $L^1[0, 1]$ (see Chapter 11). However, the question of whether such a set exists in a reflexive Banach space remains open. (Of course in such spaces all bounded, closed and convex sets are weakly compact.) This perhaps is the major remaining open question in the theory. However many other questions remain open.

We now turn to the second question raised at the outset of this discussion concerning the structure of the set of fixed points of a nonexpansive mapping T. Obviously, since such a mapping T is continuous, its fixed point set is closed. In some settings this set is also convex.

A Banach space X (or the norm $\|\cdot\|$ on X) is said to be *strictly convex* if the following implication holds for all $x, y \in X$:

$$\left.\begin{array}{c} \|x\| \leqslant 1 \\ \|y\| \leqslant 1 \\ \|x-y\| > 0 \end{array}\right\} \Rightarrow \left\|\frac{x+y}{2}\right\| < 1.$$

This is equivalent to the condition that the unit sphere (or any sphere)contains no line segments. In such a space, any three points x, z, y satisfying $\|x-z\| + \|z-y\| = \|x-y\|$ must lie on a line; specifically, if $\|x-z\| = r_1$, $\|y-z\| = r_2$, and $\|x-y\| = r = r_1 + r_2$, then $z = (r_1/r)x + (r_2/r)y$.

Lemma 3.4 *If K is a closed and convex subset of a strictly convex space X and if $T: K \to K$ is nonexpansive, then the set Fix T of fixed points of T is closed and convex.*

Proof Fix T is closed because T is continuous. Suppose $x = Tx$ and $y = Ty$, let $\lambda \in (0, 1)$, and set $z = (1-\lambda)x + \lambda y$. Then $\|x - Tz\| + \|Tz - y\| = \|Tx - Tz\| + \|Tz - Ty\| \leqslant \|x-z\| + \|z-y\| = \|x-y\| \leqslant \|x - Tz\| + \|Tz - y\|$. It follows that x, Tz, and y are colinear while $\|x-z\| = \|x - Tz\|$ and $\|y-z\| = \|y - Tz\|$. Since X is strictly convex, $z = Tz$.

The above fact does not hold in general:

Example 3.6 Assign $\mathbb{R}^2$ the equivalent norm:

$$\|(x, y)\|_\infty = \max\{|x|, |y|\},$$

and define $T: \mathbb{R}^2 \to \mathbb{R}^2$ by

$$T(x, y) = (x, |x|).$$

Then T is nonexpansive relative to $\|\cdot\|_\infty$ and Fix T is the graph $\{(x, y): y = |x|\}$, which is not convex. In fact, if $f: \mathbb{R} \to \mathbb{R}$ is any function satisfying

$$|f(t) - f(s)| \leqslant |t - s|,$$

then the graph of f is the fixed point set of the nonexpansive mapping on $(\mathbb{R}^2, \|\cdot\|_\infty)$ defined by

$$T(x, y) = (x, f(x)).$$

For example, one could take f to be the sine function. Moreover, the mapping $\lambda: \mathbb{R} \to \mathbb{R}^2$ given by $\lambda(t) = (t, f(t))$ defines an isometric parameterization of the curve $y = f(x)$, so the graph of f is a 'metric line' in $(\mathbb{R}^2, \|\cdot\|_\infty)$.

Our final example shows that the set Fix T may even be disconnected.

Example 3.7 Consider the unit ball in the space c_0 (of sequences of real numbers with limit 0 and supremum norm). Define the nonexpansive mapping $T: c_0 \to c_0$ by

$$Tx = T(x_1, x_2, \ldots) = (x_1, 1 - |x_1|, x_2, x_3, \ldots).$$

Since $\lim_{n \to \infty} x_n = 0$, if $x = Tx$ then $x_2 = x_3 = \cdots = 0$. Thus $|x_1| = 1$ and Fix T consists of the two points e^1 and $-e^1$.

The third question posed at the beginning of this chapter concerned the behavior of the sequence of iterates of a nonexpansive mapping T. We postpone a discussion of this for now, remarking only that the situation for nonexpansive mappings is much different than for contractions. In general, if $T: K \to K$ is nonexpansive and $x_0 \in K$, then $\lim_{n \to \infty} \|T^n x_0 - T^{n+1} x_0\|$ always exists, but this limit may be positive. On the other hand, if $F = (I + T)/2$ (i.e., $Fx = (x + Tx)/2$) then one may check that in each of the examples of this chapter,

$$\lim_{n \to \infty} \|F^n x - F^{n+1} x\| = 0.$$

As we shall see, this is not a coincidence.

We conclude our preliminary observations by noting that our basic fixed point results are 'separable' in nature. This is understood as follows:

Suppose X is a Banach space and K is a closed and convex subset of X which is invariant under some mapping T (which need not be nonexpansive). Select $x_0 \in K$ and consider the sequence $\{K_n\}$ of compact convex sets defined by taking $K_0 = \{x_0\}$ and setting

$$K_{n+1} = \overline{\text{conv}}(K_n \cup T(K_n)), \qquad n = 0, 1, 2, \ldots.$$

If X_n denotes the subspace of X spanned by K_n then obviously X_n is separable. Set

$$K_\infty = \overline{\bigcup_{n=0}^{\infty} K_n}$$

and

$$X_\infty = \overline{\bigcup_{n=0}^{\infty} X_n}$$

Then X_∞ is separable and K_∞ is a closed and convex subset of X_∞ which is T-invariant.

The above shows that in many cases, i.e., when the other assumptions on K are inherited by K_∞, it suffices to formulate fixed point problems in a separable setting. This is a fact we shall use later; e.g., in Chapter 15.

4

The basic fixed point theorems for nonexpansive mappings

We have already seen that some bounded, closed and convex sets K in certain Banach spaces have the property that every nonexpansive self-mapping $T: K \to K$ must have a fixed point. When this is the case we say that K has the *fixed point property* (f.p.p.) for nonexpansive mappings and, if it is clear that only nonexpansive mappings are being considered, we shall simply say that K has the fixed point property or that K has f.p.p. Also, unless otherwise specified, we shall always assume that K is nonempty, bounded, closed and convex.

The problem of determining conditions on K (or on the space X containing K) which always insure that K has the f.p.p. has its origins in four papers which appeared in 1965. In the first of these (Browder, 1965a), F. Browder proved that a bounded, closed, convex set $K \subset X$ has f.p.p. if X is a Hilbert space. Almost immediately, both Browder (1965b) and Göhde (1965) proved that the same is true if X belongs to the much wider class of 'uniformly convex' spaces (discussed in the next chapter). At the same time Kirk (1965) observed that the presence of a geometric property called 'normal structure' guarantees that $K \subset X$ has f.p.p. if X is reflexive. The concept of normal structure was introduced in 1948 by Brodskii and Milman (1948) to study fixed points of isometries, and it is a property shared by all uniformly convex spaces.

We begin with the approach of Kirk and related observations. The approaches of Browder and Göhde are also of interest for reasons we shall take up in Chapter 10.

For any subsets D, H of X, set:

$$r_u(D) = \sup\{\|u - v\| : v \in D\} \qquad (u \in X);$$

$$r_H(D) = \inf\{r_u(D) : u \in H\};$$

$$C_H(D) = \{u \in H : r_u(D) = r_H(D)\}.$$

The number $r_u(D)$ is called the *radius* of D relative to u; $r_H(D)$ and $C_H(D)$ are called, respectively, the *Chebyshev radius* and *Chebyshev center* of D relative to H. When $H = D$ the notation $r(D)$ and $C(D)$ is used for these last two entities and the phrase 'relative to H' is dropped. Note that if $u \in C_H(D)$

then $B(u; r_u(D))$ contains D, while no ball centered at any point of H with smaller radius has this property.

It is obvious that for any $u \in D$,

$$r(D) \leqslant r_u(D) \leqslant \operatorname{diam} D.$$

Moreover, it should be noted that if $r_u(D)$ is thought of as a function of u defined on a bounded, closed and convex subset D of X, then $r_u : D \to \mathbb{R}$ is both continuous and convex and $C(D)$ thus is closed and convex. This latter fact also follows from the relation:

$$C(D) = \bigcap_{\epsilon > 0} C_\epsilon(D)$$

where

$$C_\epsilon(D) = \{u \in D : r_u(D) \leqslant r(D) + \epsilon\}$$

$$= \bigcap_{x \in D} B(x; r(D) + \epsilon) \bigcap D.$$

While the sets $C_\epsilon(D)$ for $\epsilon > 0$ are all nonempty, it may be the case that $C(D) = \emptyset$. However, if D is nonempty and weakly compact, then all closed and convex subsets of D are weakly closed, hence weakly compact, and it follows that for such D, $C(D) \neq \emptyset$.

A point $u \in D$ is said to be *diametral* if $r_u(D) = \operatorname{diam} D$; if this is not the case u is said to be *nondiametral*. By looking at examples given in the previous chapter it is easy to see that a bounded, closed and convex set D may consist entirely of diametral points. Such sets are called *diametral*, and they represent a type of geometric 'abnormality'.

Example 4.1 Let M be the subset of $\mathscr{C}[0, 1]$ defined as in Example 2.1:

$$M = \{x = \{x(t)\} : 0 = x(0) \leqslant x(t) \leqslant x(1) = 1\}.$$

Consider the two norms on $\mathscr{C}[0, 1]$ obtained by defining for each $x \in \mathscr{C}[0, 1]$:

$$\|x\|_0 = \max_{0 \leqslant t \leqslant 1} \{|x(t)|\};$$

$$\|x\|_1 = \|x\|_0 + \left(\int_0^1 (x(t))^2 \mathrm{d}t \right)^{1/2}$$

With respect to $\|\cdot\|_0$, $r(M) = \operatorname{diam} M = 1$ and M is diametral, but with respect to $\|\cdot\|_1$, $\operatorname{diam} M = 2$, $r(M) = \frac{3}{2}$, and no point of M is diametral. In the first case $C(M) = M$ and in the second $C(M) = \emptyset$.

Definition 4.1 (Brodskii and Milman, 1948) A convex subset K of X is said to have *normal structure* if each bounded, convex subset S of K with diam $S > 0$ contains a nondiametral point.

Thus sets with normal structure have no convex subsets S which consist entirely of diametral points except singletons; i.e.,

$$\text{diam } S > 0 \Rightarrow r(S) < \text{diam } S.$$

Using a straightforward transfinite induction argument, Brodskii and Milman (1948) showed that a weakly compact convex set K which has normal structure always contains a point $\bar{x}$ which is fixed under every isometry which maps K onto K. The fact that all compact convex sets have normal structure, is an immediate consequence of the following lemma. As we shall see later, the same is true for all bounded, convex subsets of uniformly convex spaces. In fact, these latter sets satisfy a stronger property we discuss later in this chapter called 'uniform normal structure'. (See Remark 5.1.)

The following characterization of normal structure is also due to Brodskii and Milman. It involves sequences of the following type: A bounded sequence $\{x_n\}$ in a Banach space X is said to be *diametral* if it is not eventually constant and if

$$\lim_{n \to \infty} \text{dist}(x_{n+1}, \text{conv}\{x_1, x_2, \ldots, x_n\}) = \text{diam}\{x_1, x_2, \ldots\}.$$

Lemma 4.1 *A bounded convex subset K of a Banach space has normal structure if and only if it does not contain a diametral sequence.*

Proof If K contains a diametral sequence $\{x_n\}$ then $S = \text{conv}\{x_1, x_2, \ldots\}$ is diametral. Conversely, assume that K contains a nontrivial, diametral, convex set S. Let $d = \text{diam } S$ and choose $\epsilon \in (0, d)$. Starting with $x_1 \in S$ it is possible to define a sequence $\{x_n\}$ in S inductively which satisfies

$$\| y_{n-1} - x_{n+1} \| > d - \frac{\epsilon}{n^2}, \qquad \text{where } y_{n-1} = \sum_{i=1}^{n} \frac{x_i}{n}.$$

Now take $x \in \text{conv}\{x_1, \ldots, x_n\}$, say $x = \Sigma_{j=1}^{n} \alpha_j x_j$ where $\alpha_j \geq 0$ and $\Sigma_{j=1}^{n} \alpha_j = 1$. If $\alpha = \alpha_p = \max\{\alpha_1, \ldots, \alpha_n\}$, then

$$y_{n-1} = \frac{x}{n\alpha} + \sum_{j=1}^{n} \left(\frac{1}{n} - \frac{\alpha_j}{n\alpha} \right) x_j.$$

Note that

$$\frac{1}{n\alpha} + \sum_{j=1}^{n} \left(\frac{1}{n} - \frac{\alpha_j}{n\alpha} \right) = 1 \qquad \text{and} \qquad \frac{1}{n} - \frac{\alpha_j}{n\alpha} \geq 0,$$

and consequently

$$d - \frac{\epsilon}{n^2} < \| y_{n-1} - x_{n+1} \|$$

$$\leqslant \frac{1}{n\alpha} \| x - x_{n+1} \| + \sum_{j \neq p} \left(\frac{1}{n} - \frac{\alpha_j}{n\alpha} \right) \| x_j - x_{n+1} \|$$

$$\leqslant \frac{1}{n\alpha} \| x - x_{n+1} \| + \left(1 - \frac{1}{n\alpha} \right) d.$$

Hence

$$\| x - x_{n+1} \| \geqslant \left(\frac{d}{n\alpha} - \frac{\epsilon}{n^2} \right) n\alpha = d - \frac{\epsilon\alpha}{n} \geqslant d - \frac{\epsilon}{n}.$$

Since $x \in \text{conv}\{x_1, \ldots, x_n\}$ and ϵ is arbitrary, the conclusion follows.

The above fact may be easily applied to identify reflexive spaces which do not have normal structure. Since any subsequence of a diametral sequence is also diametral and since the weak limit of a sequence belongs to its convex envelope, if a space X is reflexive and does not have normal structure then X contains a sequence $\{x_n\}$ for which:

$$w\text{-}\lim_{n \to \infty} x_n = 0;$$

$$\lim_{n \to \infty} \| x_n \| = 1;$$

$$\text{diam}\,(\{x_n\}) \leqslant 1.$$

The following is widely regarded as the fundamental existence theorem for nonexpansive mappings. Since bounded, closed, convex subsets of reflexive Banach spaces are always weakly compact, this theorem applies to all such spaces.

Theorem 4.1 *Let K be a nonempty, weakly compact, convex subset of a Banach space, and suppose K has normal structure. Then every nonexpansive mapping $T: K \to K$ has a fixed point.*

Proof 1 As noted previously, we may assume K is minimal invariant under T, and this implies

$$\overline{\text{conv}}\, T(K) = K.$$

Let $u \in C(K)$; i.e., $r_u(K) = r(K)$. Since $\|Tu - Tv\| \leqslant \|u - v\| \leqslant r(K)$ for all $v \in K$,

$$T(K) \subset B(Tu, r(K)).$$

Consequently,

$$K = \overline{\text{conv}}\, T(K) \subset B(Tu; r(K))$$

showing that $r_{Tu}(K) = r(K)$; thus $Tu \in C(K)$. This proves that $C(K)$ is T-invariant contradicting (in view of normal structure) the minimality of K unless diam $K = 0$, an alternative which implies K consists of a single point which is fixed under T.

The above is a reformulation of the original proof of Kirk (1965). Its initial step – the existence of a minimal T-invariant subset of K – is crucial and involves an application of Zorn's Lemma. More 'constructive' proofs exist. The one we give next relies on the following lemma (essentially due to Gillespie and Williams, 1979).

Lemma 4.2 *Let K be a nonempty, bounded, closed and convex subset of X and let $T: K \rightarrow K$ be nonexpansive. Then there exists a nonempty, closed, convex and T-invariant subset D of K such that*

$$\text{diam } D \leqslant \tfrac{1}{2}(\text{diam } K + r(K)). \tag{4.1}$$

Proof We construct D as follows: Let $\rho = \tfrac{1}{2}(\text{diam } K + r(K))$ and let

$$C = \{z \in K : r_z(K) \leqslant \rho\} = \left(\bigcap_{x \in K} B(x; \rho)\right) \cap K.$$

If $\rho = \text{diam } K$ then K is diametral and we may take $D = K$. Otherwise $r(K) < \rho < \text{diam } K$ and C is a nonempty, proper, closed, convex subset of K.
    Now, as in Theorem 3.1, let $\mathcal{M}$ denote the family of all nonempty, closed, convex and T-invariant subsets of K, and set

$$\mathcal{F} = \{H \in \mathcal{M} : C \subset H\} \text{ and } L = \bigcap \mathcal{F}.$$

Note that $\mathcal{F} \neq \emptyset$ since $K \in \mathcal{F}$; hence $L \supset C \neq \emptyset$. Moreover $L \in \mathcal{M}$. Let $A = C \cup T(L)$. Then $A \subset L$; hence $\overline{\text{conv}}\, A \subset L$. Also,

$$T(\text{conv } A) \subset T(L) \subset A \subset \text{conv } A.$$

This proves $\overline{\text{conv}}\, A \in \mathcal{F}$, so by the definition of L, $\overline{\text{conv}}\, A = L$.
    Now let

$$D = \{x \in L : r_x(L) \leqslant \rho\} = \left(\bigcap_{x \in L} B(x; \rho)\right) \cap L.$$

Then $C \subset D$, so $D \neq \emptyset$. Also, if $x \in D$, then $Tx \in L$ and for each $y \in L$, $\|Tx - Ty\| \leqslant \|x - y\| \leqslant \rho$. Furthermore, if $z \in C$, $\|Tx - z\| \leqslant \rho$ (because $D \subset K \subset B(z; \rho)$). This proves that $A = C \cup T(L) \subset B(Tx; \rho)$, i.e., $T: D \to D$. Since D is closed and convex by definition, it follows that $D \in \mathcal{M}$ and, since diam $D \leqslant \rho$, the proof is complete.

Now we present a more constructive proof of Theorem 4.1 in the sense that it avoids the use of Zorn's Lemma. (Another constructive proof, in a uniformly convex setting, is given in Goebel, 1969.)

***Proof* 2** Define $\mathcal{M}$ as in the lemma and define $\delta_0: \mathcal{M} \to \mathbb{R}^+$ by

$$\delta_0(D) = \inf\{\text{diam } F: F \in \mathcal{M}, F \subset D\}.$$

Set $D_1 = K$, and with $D_1, \dots, D_n$ given, select $D_{n+1} \in \mathcal{M}$ so that $D_{n+1} \subset D_n$ and

$$\text{diam } D_{n+1} \leqslant \delta_0(D_n) + 1/n.$$

Let

$$C = \bigcap_{n=1}^{\infty} D_n.$$

By weak compactness, $C \neq \emptyset$; hence $C \in \mathcal{M}$. Now let D be a subset of C constructed as in the lemma so that $D \in \mathcal{M}$ and

$$\text{diam } D \leqslant \tfrac{1}{2}(\text{diam } C + r(C)).$$

Then

$$\text{diam } C - \frac{1}{n} \leqslant \text{diam } D_{n+1} - \frac{1}{n} \leqslant \delta_0(D_n) \leqslant \text{diam } D \leqslant \tfrac{1}{2}(\text{diam } C + r(C)).$$

Letting $n \to \infty$, diam $C \leqslant \tfrac{1}{2}(\text{diam } C + r(C))$ yielding diam $C = 0$ so, as in Proof 1, C is a singleton which is invariant under T.

There is another proof of Theorem 4.1 which is of interest in part because it relies on the following 1905 theorem of Zermelo which avoids the Axiom of Choice (see Dunford and Schwartz, 1957, p. 5). This proof is also 'constructive' in the sense of Proof 2.

Theorem 4.2 (Zermelo) *Let $(E, \leqslant)$ be a nonempty, partially ordered set with the properties:*

(a) *$a \leqslant b$ and $b \leqslant a \Rightarrow a = b$.*
(b) *Each chain (totally ordered subset) in E has a least upper bound.*

Suppose $f: E \to E$ satisfies $fx \geqslant x$, $x \in E$. Then for each $x \in E$ there exists a point $z \geqslant x$ depending constructively on x for which $fz = z$.

The following proof is a minor variant of one given by Fuchssteiner (1977). Note that two applications of Zermelo's Theorem are required.

Proof 3 Define $\mathcal{M}$ as in Proof 1.

Step 1 Fix $Y \in \mathcal{M}$ and let $\mathcal{M}(Y) = \{ D \in \mathcal{M} : D \subset Y \}$. Now order $\mathcal{M}(Y)$ by taking $D_2 \leqslant D_1 \Leftrightarrow D_1 \subset D_2$ and define $f_1 : \mathcal{M}(Y) \to \mathcal{M}(Y)$ by

$$f_1(D) = \overline{\text{conv}}(T(D)), \, D \in \mathcal{M}(Y).$$

Since $T(D) \subset D$, $\overline{\text{conv}}\, T(D) \subset D$; hence $f_1(D) \geqslant D$. Also, by weak compactness every chain in $(\mathcal{M}(Y), \leqslant)$ has a least upper bound – the intersection of all its members. Since $Y \in \mathcal{M}(Y)$, Zermelo's Theorem yields a set $Y_0 \in \mathcal{M}(Y)$, depending on Y, such that $f_1(Y_0) = Y_0$, i.e.,

$$\text{conv}(T(Y_0)) = Y_0.$$

It is now possible to prove $C(Y_0) \in \mathcal{M}$ as in Proof 1.

Step 2 Now define $f_2 : \mathcal{M} \to \mathcal{M}$ by

$$f_2(Y) = C(Y_0), \qquad Y \in \mathcal{M}.$$

Order $\mathcal{M}$ by $Y_2 \geqslant Y_1 \Leftrightarrow Y_2 \subset Y_1$. For each $Y \in \mathcal{M}$, $C(Y_0) \subset Y_0 \subset Y$; thus $f_2(Y) \geqslant Y$. Since every chain in $\mathcal{M}$ also has a least upper bound, a second application of Zermelo's Theorem yields $Y \in \mathcal{M}$ such that $f_2(Y) = Y$. Thus

$$C(Y_0) = Y = Y_0.$$

Since K has normal structure, this implies diam $Y_0 = 0$, so $Y_0 \in \mathcal{M}$ is a singleton, i.e. a fixed point of T.

As we have seen, the Chebyshev center $C(D)$ of a set $D \subset X$ is characterized by

$$C(D) = \bigcap_{\epsilon > 0} C_\epsilon(D)$$

where

$$C_\epsilon(D) = \{ u^\epsilon D : u \in \bigcap_{x \in D} B(x; r(D) + \epsilon) \}.$$

It is obvious that $C(D) \neq \emptyset$ if D is compact in some topology for which the balls in X are also compact. In particular, if X is a Banach space which is the dual of another space and if D is weak* compact, then $C(D) \neq \emptyset$. This fact suggests the following:

Definition 4.2 A convex set D in a dual Banach space X is said to have *weak* normal structure* if each bounded weak*-closed convex subset S of D with diam $S > 0$ contains a nondiametral point.

Example 4.2 Weak*, compact, convex subsets of l^1 (as the dual of c_0) have weak* normal structure.

To see this let $K \subset l^1$ be weak*, compact and convex and let $S \subset K$ be weak* closed and convex with diam $S > 0$. If S is compact then it follows immediately from Lemma 4.1 that S has normal structure and hence contains a nondiametral point. So we may suppose S contains a weak* convergent sequence $\{x_n\}$ such that weak* $\lim_{n \to \infty} x_n = x \in S$ but such that $\{x_n\}$ does not converge strongly to x.

For each $y \in S$, let

$$r(y) = \limsup_{n \to \infty} \|x_n - y\|.$$

Then by Lemma 3.2,

$$r(y) = r(x) + \|y - x\|,$$

and this implies the existence of a nondiametral point in S since

$$\|x - y\| \leq r(y) - r(x) \leq \text{diam } S - r(x)$$

implies

$$r_x(S) \leq \text{diam } S - r(x),$$

with $r(x) > 0$.

The following analogue of Theorem 4.1 holds in all dual spaces and, in view of the above example, applies to $X = l^1$.

Theorem 4.3 *Let K be a nonempty weak*, compact, convex subset of a dual Banach space X, and suppose K has weak* normal structure. Then every nonexpansive mapping $T: K \to K$ has a fixed point.*

The proof of the above is identical to that of Theorem 4.1 except that $\overline{\text{conv}}\, T(K)$ must be replaced with $\overline{\text{conv}}^*\, T(K)$, i.e., the intersection of all weak*, closed, convex sets containing $T(K)$.

While the second proof of Theorem 4.1 is, in a logical sense, more constructive than the original one it does not provide an algorithmic approach leading to a fixed point. This changes if the space has even better geometric structure.

Definition 4.3 A nonempty, bounded, convex set K in X is said to have *uniform normal structure* if there exists a constant $k \in (0, 1)$ such that

$$r(D) \leqslant k \text{ diam } D$$

for any closed convex subset D of K. The space X is said to have uniform normal structure if each of its nonempty, convex subsets has this property.

Uniform normal structure is obviously stronger than normal structure and as we shall see in the next chapter it mimics the 'normal structure' behavior of uniformly convex spaces. Also, nonreflexive spaces exist which have normal structure. This contrasts with the following (cf., Maluta, 1989):

Theorem 4.4 *If a Banach space X has uniform normal structure, then it is reflexive.*

Proof We first show that any descending sequence $\{K_n^0\}_{n=1}^{\infty}$ of nonempty, bounded, closed, convex subsets of X has nonempty intersection. Let $k_0 = \sup\{r(C)/\text{diam } C\}$ where the supremum is taken over all nonempty, closed, convex subsets C of X with diam $C > 0$. Since $k_0 < 1$ it is possible to select $k \in (k_0, 1)$. For C as above, set

$$A(C) = \{x \in C : \|x - y\| \leqslant k \text{ diam } C \text{ for each } y \in C\}.$$

Obviously $A(C)$ is also nonempty, closed and convex, and, moreover, $A(C)$ is a proper subset of C.

Now set

$$K_1^1 = \overline{\text{conv}} \bigcup_{i=1}^{\infty} A(K_i^0);$$

$$K_2^1 = \overline{\text{conv}} \bigcup_{i=2}^{\infty} A(K_i^0);$$

$$\vdots$$

$$K_n^1 = \overline{\text{conv}} \bigcup_{i=n}^{\infty} A(K_i^0).$$

$$\vdots$$

We claim diam $K_n^1 \leqslant k$ diam K_n^0. To see this, let $x, y \in \bigcup_{i=n}^{\infty} A(K_i^0)$. Then $x \in A(K_p^0)$ and $y \in A(K_q^0)$ with, say, $n \leqslant p \leqslant q$. Since $K_q^0 \subset K_p^0$, the definition of

$A(K_p^0)$ implies

$$\|x-y\| \leqslant k \operatorname{diam} K_p^0 \leqslant k \operatorname{diam} K_n^0,$$

and our claim follows.

By repeating the above construction it is possible to obtain the following:

$$K_1^0 \supset K_2^0 \supset \cdots \supset K_n^0 \supset \cdots$$
$$\cup \quad \cup$$
$$K_1^1 \supset K_2^1 \supset \cdots \supset K_n^1 \supset \cdots$$
$$\cup \quad \cup \qquad \cup$$
$$\vdots \quad \vdots \qquad \vdots$$
$$\cup \quad \cup \qquad \cup$$
$$K_1^n \supset K_2^n \supset \cdots \supset K_n^n \supset \cdots$$

Since $\operatorname{diam} K_i^n \leqslant k^n \operatorname{diam} K_1^0 \to 0$ as $n \to \infty$, the diagonal sequence $\{K_{n+1}^n\}_{n=1}^\infty$ has nonempty intersection by Cantor's Theorem. It is now possible to invoke theorems of Smulian and Eberlein – Smulian to conclude that X is reflexive (see Dunford and Schwartz, 1957, p. 433).

If X has uniform normal structure and if K is a nonempty, bounded, closed, convex subset of X, the construction of Lemma 4.2 yields sets

$$K = D_1, D_2, \ldots, \qquad D_i \in \mathcal{M} \ (\mathcal{M} \text{ as in Lemma 4.2})$$

satisfying

$$\operatorname{diam} D_{n+1} \leqslant \tfrac{1}{2}(\operatorname{diam} D_n + r(D_n))$$
$$\leqslant \tfrac{1}{2}(1+k) \operatorname{diam} D_n$$
$$\leqslant \cdots$$
$$\leqslant \left(\frac{1+k}{2}\right)^n \operatorname{diam} K.$$

Obviously $\lim_{n \to \infty} \operatorname{diam} D_n = 0$, so $\bigcap_{n=1}^\infty D_n$ consists of a single point which is fixed under T.

The concept of normal structure has been widely studied and has many natural variations (see, e.g., Landes, 1984 and 1986, and the surveys by Kirk (1981a, 1983) and Swaminathan 1983). We concentrate on a variation connected with the notion of hyperconvexity which has fixed point theoretic implications in the classical spaces $\mathscr{C}(K)$ and $L^\infty(\Omega, \Sigma, \mu)$. These spaces in general lack both normal and weak* normal structure.

Definition 4.4 A metric space (M, ρ) is said to be *hyperconvex* if any family

$\{B(x_\alpha; r_\alpha)\}$ of closed balls in M satisfying $\rho(x_\alpha, x_\beta) \leqslant r_\alpha + r_\beta$ has nonempty intersection.

The real line is hyperconvex (a fact commonly known as Helly's Theorem) as are the spaces $\mathbb{R}^n$ furnished with the l^∞-norm ($\|(x_1, \ldots, x_n)\|_\infty = \max\{|x_i|: 1 \leqslant i \leqslant n\}$). In fact, the general structure of Banach spaces whose norms generate hyperconvex metrics is well known. A Banach space X is hyperconvex if and only if it is *isometrically isomorphic to* $\mathscr{C}(K)$ *where* K *is a compact extremally disconnected* (the closure of each open set is itself open) *Hausdorff topological space* (Aronszajn and Panitchpakdi, 1956; Nachbin, 1950; Kelley, 1952)). Such sets K are called *Stonian* in honor of A.H. Stone, who proved that every complete Boolean algebra is isomorphic to the Boolean algebra of all open – closed subsets of such a space.

The Stonian structure of K has strong implications: The spaces $\mathscr{C}(K)$ are complete lattices with respect to their natural ordering, and they have both the Hahn–Banach extension property and the 1-projection property (see, e.g., Lacey, 1974).

Hyperconvex spaces are of interest in a classical sense because of the L^∞ spaces. Let (Ω, Σ, μ) be a measure space and let $\mathscr{N}$ be the ideal of null sets in Σ. Then the space $L^\infty(\Omega, \Sigma, \mu)$ of all μ-measurable real-valued functions f on Ω for which

$$\|f\|_{L^\infty} = \text{ess sup}\{|f(t)|: t \in \Omega\}$$

$$= \inf_{E \in \mathscr{N}} \sup_{\Omega \setminus E} |f| < \infty$$

is isometrically isomorphic to a space $\mathscr{C}(K)$ with K Stonian. Thus L^∞ is itself hyperconvex. (The space K is unique up to homeomorphism and may be described in several ways. For example, K may be taken to be a space consisting of certain extremal points of the unit ball in the space $(L^\infty)^*$ (of multiplicative functionals).) (For details, see Dunford and Schwartz, 1957; also Schaefer, 1974; Lacey, 1974.)

We now return to hyperconvexity in a metric setting. Since the notions of the radius of a set relative to a point, Chebyshev radius, and Chebyshev center, which were introduced in a Banach space setting at the beginning of this chapter, are purely metric in nature we shall adopt their metric analogues and notation accordingly.

Now let (M, ρ) be a hyperconvex metric space. It is easy to verify that the hyperconvexity of M implies its completeness. Also, if $J = \bigcap_{\alpha \in A} B(x_\alpha; r_\alpha)$ is the intersection of closed balls in M, then (J, ρ) is itself a hyperconvex metric space. Let $\mathscr{J}$ denote the family of all subsets of M of this type and note in particular that $\mathscr{J}$ is compact in the sense that any subfamily of $\mathscr{J}$ with the finite intersection property has nonempty intersection.

Now take $D \in \mathscr{J}$ with $d = \operatorname{diam} D > 0$. For any $u, v \in D$, $\rho(u, v) \leqslant d = \frac{1}{2}d + \frac{1}{2}d$. Thus by hyperconvexity, the Chebyshev radius of D relative to M (or to D) is equal to $\frac{1}{2}d$, and the Chebyshev centers are given by

$$C_M(D) = \bigcap \{B(u; \tfrac{1}{2}d): u \in D\},$$
$$C_D(D) = \bigcap \{B(u; \tfrac{1}{2}d): u \in D\} \cap D$$
$$= C_M(D) \cap D.$$

Both $C_M(D)$ and $C_D(D)$ are nonempty and belong to $\mathscr{J}$. Moreover, for any $z \in C_D(D)$,

$$B(z; \tfrac{1}{2}d) \supset C_M(D)$$

and $\operatorname{diam} C_D(D) \leqslant d/2$. (However, it may be the case that $\operatorname{diam} C_M(D) > \frac{1}{2}d$ (even $= d$). Appropriate examples can be constructed even in the plane l_2^∞.)

An operation similar to that of convex closure can be defined more abstractly as well. For any set $S \subset M$, set

$$\operatorname{cov} S = \bigcap \{C: C \in \mathscr{J}, C \supset S\}.$$

Obviously $\operatorname{cov} S \in \mathscr{J}$ and $S \subset \operatorname{cov} S$.

Since the situation described above closely resembles the development leading to Theorem 4.1, it is not surprising that we have the following theorem (implicit in Soardi, 1979).

Theorem 4.5 *Suppose (M, ρ) is a bounded, hyperconvex, metric space. Then every nonexpansive mapping $T: M \to M$ has a fixed point.*

Proof As in Theorem 4.1 (Proof 1) consider the family

$$\mathscr{M} = \{D \in \mathscr{J}; T: D \to D\}.$$

Since $M \in \mathscr{M}$, $\mathscr{M} \neq \emptyset$, so by compactness of $\mathscr{J}$ and Zorn's Lemma, $\mathscr{M}$ has a minimal element M_0. Since $T(M_0) \subset M_0$, $\operatorname{cov}(T(M_0)) \subset M_0$ from which

$$T(\operatorname{cov}(T(M_0))) \subset T(M_0) \subset \operatorname{cov}(T(M_0)).$$

Thus $\operatorname{cov}(T(M_0)) \in \mathscr{M}$, so by minimality

$$\operatorname{cov}(T(M_0)) = M_0.$$

Now suppose $\operatorname{diam} M_0 = d > 0$ and consider the Chebyshev center

$$C_{M_0}(M_0) = \cap \{B(u; \tfrac{1}{2}d): u \in M_0\} \cap M_0.$$

If $u \in M_0$ and $v \in C_{M_0}(M_0)$ then $\rho(Tu, Tv) \leqslant \rho(u, v) \leqslant \frac{1}{2}d$; thus $T(M_0) \subset$

$B(Tv; \frac{1}{2}d)$ yielding $M_0 = \text{cov}(T(M_0)) \subset B(Tv; \frac{1}{2}d)$. Therefore $Tv \in C_{M_0}(M_0)$, i.e.,

$$T: C_{M_0}(M_0) \to C_{M_0}(M_0)$$

so $C_{M_0}(M_0) \in \mathcal{M}$. However $C_{M_0}(M_0) \neq M_0$ because $\text{diam } C_{M_0}(M_0) \leqslant \frac{1}{2}d$. This contradicts the minimality of M_0, so $d = 0$; thus $M_0 = \{z\}$ and $Tz = z$.

Remark 4.1 It should be noted that in the L^∞ spaces the sets $\mathcal{J}$ correspond to order intervals (sets of the form $\{f : h \leqslant f \leqslant g\}$ for fixed $h \leqslant g$ in L^∞). Thus Theorem 4.4 implies that order intervals in L^∞ have the f.p.p. for nonexpansive mappings – a fact first noted by Soardi (1979) and Sine (1979). For related results see Borwein and Sims (1984).

Remark 4.2 Theorem 4.1 has a more abstract formulation, due to Penot (1979). Suppose the metric space (M, ρ) possesses a *convexity structure* $\mathscr{L} \subset 2^M$ in the sense:

(a) $\emptyset, M \in \mathscr{L}$;
(b) $\{x\} \in \mathscr{L}$;
(c) $\mathscr{F} \subset \mathscr{L} \Rightarrow \bigcap \{F : F \in \mathscr{F}\} \in \mathscr{L}$.

For $S \subset M$, the set

$$\text{cov}(S) = \bigcap \{C : C \in \mathscr{L}, C \supset S\}$$

is called the $\mathscr{L}$-*hull* of S. A convexity structure is said to be *normal* if for each $D \in \mathscr{L}$, $\text{diam } D > 0 \Rightarrow r(D) < \text{diam } D$, and *sequentially compact* if each decreasing sequence of nonempty subsets of M has nonempty intersection.

The following is proved in Kirk (1981b). The line of argument is identical with that of Proof 2 of Theorem 4.1.

Theorem 4.6 *Let (M, ρ) be nonempty and bounded, and suppose M possesses a convexity structure $\mathscr{L}$ which is sequentially compact, normal and contains the closed balls of M. Then every nonexpansive mapping $T: M \to M$ has a fixed point.*

Theorem 4.6 includes Theorems 4.1, 4.3, and 4.5 as special cases. An application which requires the sequential version of compactness is given in Khamsi, Kozlowski and Reich, (1989).

Remark 4.3 Both Theorems 4.1 and 4.5 have been extended to abelian (and in the case of Theorem 4.1, left reversible) semigroups of nonexpansive mappings. For these and related results we refer the reader to Bruck (1974), Lim (1974a), Khamsi (1987), Lin and Sine (1988), and Baillon 1988.

Remark 4.4 Soardi (1979) proved a version of Theorem 4.5 which applies to wider classes of spaces, including the complex Banach lattices (see Kirk, 1983). In abstract terms Soardi's result may be formulated as follows.

Let (M, ρ) be a metric space and let $\mathscr{J}$ denote the family of closed balls in M. M is said to have *uniformly relative normal structure* if there exists $c \in (0, 1)$ such that for each $D \in \mathscr{J}$ there exists $z \in M$ satisfying

(a) $D \subset B(z; c \text{ diam } D)$; and
(b) $u \in M$ and $D \subset B(u; c \text{ diam } D) \Rightarrow \rho(u, z) \leqslant c \text{ diam } D$.

As seen in the proof of Theorem 4.5, hyperconvex metric spaces always satisfy (a) and (b) for $c = \frac{1}{2}$. A suitable modification of Soardi's proof yields the abstract version of his result:

Theorem 4.7 *Suppose* (M, ρ) *is a bounded metric space which has uniformly relative normal structure and suppose the family* $\mathscr{J}$ *of nonempty intersections of closed balls in* M *forms a compact convexity structure. Then every nonexpansive mapping* $T: M \to M$ *has a fixed point.*

5

Scaling the convexity of the unit ball

The class of Banach spaces having normal structure is in some sense 'nice' since no space in this class contains nontrivial bounded convex sets consisting entirely of diametral points. There are other ways to scale the geometry of Banach spaces and one of the most natural of these involves the nature of the convexity of the unit sphere.

As noted previously the strictly convex spaces are those spaces for which points x, y satisfy the implication:

$$\left.\begin{array}{l} \|x\| \leqslant 1 \\ \|y\| \leqslant 1 \\ \|x-y\| > 0 \end{array}\right\} \Rightarrow \left\|\frac{x+y}{2}\right\| < 1.$$

A strengthening of the above condition was introduced by Clarkson in 1936.

Definition 5.1 A Banach space X is said to be *uniformly convex* if for each $\epsilon \in (0, 2]$ there exists $\delta > 0$ such that for $x, y \in X$,

$$\left.\begin{array}{l} \|x\| \leqslant 1 \\ \|y\| \leqslant 1 \\ \|x-y\| > \epsilon \end{array}\right\} \Rightarrow \left\|\frac{x+y}{2}\right\| \leqslant \delta.$$

Obviously uniformly convex spaces are strictly convex. Moreover, it is not difficult to see that the two concepts are equivalent in finite dimensional spaces (since balls in such spaces are compact). In general the concepts differ:

Example 5.1 Fix $\mu > 0$ and let $\mathscr{C}[0, 1]$ be given the norm $\|\cdot\|_\mu$ defined as follows.

$$\|x\|_\mu = \|x\|_0 + \mu \left(\int_0^1 x^2(t)\, dt \right)^{1/2},$$

51

where $\|\cdot\|_0$ is the usual supremum norm. Then

$$\|x\|_0 \leqslant \|x\|_\mu \leqslant (1+\mu)\|x\|_0, \qquad x \in \mathscr{C}[0,1],$$

and the two norms are equivalent with $\|\cdot\|_\mu$ near $\|\cdot\|_0$ for small μ. However $(\mathscr{C}[0,1], \|\cdot\|_0)$ is not strictly convex while for any $\mu>0$, $(\mathscr{C}[0,1], \|\cdot\|_\mu)$ is. On the other hand, it is easy to see that for any $\epsilon \in (0,2)$ there exist functions $x, y \in \mathscr{C}[0,1]$ with $\|x\|_\mu = \|y\|_\mu = 1$, $\|x-y\| = \epsilon$, and $\|(x+y)/2\|$ arbitrarily near 1. Thus $(\mathscr{C}[0,1], \|\cdot\|_\mu)$ is not uniformly convex.

Example 5.2 Let $\mu>0$ and let $c_0 = c_0(\mathbb{N})$ be given the norm $\|\cdot\|_\mu$ defined for $x = \{x_i\} \in c_0$ by

$$\|x\|_\mu = \|x\|_{c_0} + \mu\left(\sum_{i=1}^\infty \left(\frac{x_i}{i}\right)^2\right)^{1/2}.$$

As in Example 5.1, the spaces $(c_0, \|\cdot\|_\mu)$ for $\mu>0$ are strictly convex but not uniformly convex, while c_0 with its usual norm is not strictly convex.

The following concept is useful in studying such geometric properties of Banach spaces more systematically.

Definition 5.2 The *modulus of convexity* of a Banach space X is the function $\delta_X: [0,2] \to [0,1]$ defined by

$$\delta_X(\epsilon) = \inf\left\{1 - \left\|\frac{x+y}{2}\right\| : \|x\| \leqslant 1, \|y\| \leqslant 1, \|x-y\| \geqslant \epsilon\right\}. \qquad (5.1)$$

Note that for any $\epsilon>0$ the number of $\delta_X(\epsilon)$ is the largest number for which the following implication always holds: For $x, y \in X$,

$$\left.\begin{array}{r}\|x\| \leqslant 1 \\ \|y\| \leqslant 1 \\ \|x-y\| \geqslant \epsilon\end{array}\right\} \Rightarrow \left\|\frac{x+y}{2}\right\| \leqslant 1 - \delta_X(\epsilon). \qquad (5.2)$$

When working with a given space X we shall drop the subscript X and simply write δ or, for example, we may write δ_1, δ_2 to distinguish between the moduli of convexity of $(X, \|\cdot\|_1)$, $(X, \|\cdot\|_2)$.

Obviously a space X is uniformly convex if and only if its modulus of convexity satisfies $\delta(\epsilon)>0$ for $\epsilon>0$, and it is an easy exercise to show that if

$\{x_n\}$ and $\{y_n\}$ are sequences in a uniformly convex space, then the condition

$$\lim_{n\to\infty} \|x_n\| = \lim_{n\to\infty} \|y_n\| = \lim_{n\to\infty} \tfrac{1}{2}\|x_n + y_n\| = 1$$

implies $\lim_{n\to\infty} \|x_n - y_n\| = 0$.

For later reference we note that (5.2) has the following equivalent formulation. For $x, y, p \in X, R > 0$, and $r \in [0, 2R]$,

$$\left.\begin{array}{r}\|x - p\| \leqslant R \\ \|y - p\| \leqslant R \\ \|x - y\| \geqslant r\end{array}\right\} \Rightarrow \|p - \tfrac{1}{2}(x + y)\| \leqslant \left(1 - \delta\left(\frac{r}{R}\right)\right)R. \qquad (5.3)$$

Remark 5.1 Let D be a bounded, convex subset of a uniformly convex Banach space X with diam $D = d > 0$. Then if $x, y \in D$ satisfy $\|x - y\| \geqslant d/2$ and if $m = \tfrac{1}{2}(x + y)$, then for any $z \in D$,

$$r_m(D) \leqslant \|z - m\| \leqslant (1 - \delta(\tfrac{1}{2}))d.$$

Thus we conclude that X has uniform normal structure.

Two additional observations are sometimes useful. The first is that the function $\delta: [0, 2] \to [0, 1]$ of (5.1) can also be defined (equivalently) as follows:

$$\delta(\epsilon) = \inf\left\{1 - \left\|\frac{x + y}{2}\right\| : \|x\| = \|y\| = 1, \|x - y\| \geqslant \epsilon\right\}. \qquad (5.4)$$

Also, the modulus of convexity has 'two-dimensional character' in the sense that for $\epsilon \in [0, 2]$

$$\delta_X(\epsilon) = \inf \delta_E(\epsilon), \qquad (5.5)$$

where the infimum is taken over all two-dimensional subspaces of X. This second observation is obvious while the first requires a simple, yet slightly technical proof. (For example, see Lindenstrauss and Tzafriri, 1977, Vol. II, p. 60.)

The following definition identifies a feature of the modulus of convexity which has important implications in geometric fixed point theory.

Definition 5.3 The *characteristic* (or *coefficient*) of convexity of a Banach space X is the number

$$\epsilon_0 = \epsilon_0(X) = \sup\{\epsilon \geqslant 0: \delta(\epsilon) = 0\}.$$

The geometrical significance of ϵ_0 is clear. It bounds the lengths of segments which lie either on or arbitrarily near the unit sphere of X.

It is obvious that the modulus of convexity of a given space X is nondecreasing. More can be said:

Lemma 5.1 *Let X be a Banach space with modulus of convexity δ and characteristic of convexity ϵ_0. Then δ is continuous on $[0, 2)$ and strictly increasing on $[\epsilon_0, 2]$.*

Proof We begin by assuming X is two-dimensional; thus its unit ball is a bounded, closed, convex set in $\mathbb{R}^2$ which is symmetric about 0. Let $u, v \in X$ be linearly independent with $\|u\| = \|v\| = 1$, and define

$$\delta_{u,v}(\epsilon) = \inf\left\{ 1 - \left\| \frac{x+y}{2} \right\| : \|x\| \leqslant 1, \|y\| \leqslant 1, \|x-y\| \geqslant \epsilon; x - y = \lambda u, x + y = \mu v \right\}.$$

It is easy to see that $\delta_{u,v}$ is a convex function. Moreover,

$$\delta(\epsilon) = \inf\{\delta_{u,v}(\epsilon): \|u\| = 1, \|v\| = 1, u \neq \pm v\}. \tag{5.6}$$

Now let $a \in (0, 2)$ and $\epsilon_1, \epsilon_2 \in [0, a]$. The convexity of $\delta_{u,v}$ implies

$$\frac{\delta_{u,v}(\epsilon_2) - \delta_{u,v}(\epsilon_1)}{\epsilon_2 - \epsilon_1} \leqslant \frac{\delta_{u,v}(2) - \delta_{u,v}(a)}{2 - a} \leqslant \frac{1}{2-a}.$$

This shows that the directional moduli $\delta_{u,v}$ form an equicontinuous family on the interval $[0, a]$. In view of (5.5) and (5.6), for an arbitrary Banach space X and $0 \leqslant \epsilon_1 \leqslant \epsilon_2 \leqslant a < 2$,

$$\delta_X(\epsilon_2) - \delta_X(\epsilon_1) \leqslant \frac{\epsilon_2 - \epsilon_1}{2-a}.$$

Strict monotonicity of δ_X on $[\epsilon_0, 2]$ follows easily from the fact that δ_X is the infimum of the nondecreasing, convex functions $\delta_{u,v}$. (We remark that this fact does not imply convexity of δ_X; indeed, there exist two-dimensional spaces with nonconvex moduli of convexity; see Example 5.8.)

Obviously the modulus of convexity of a Banach space is an isometric invariant; in fact, if $(X, \|\cdot\|_X)$ and $(Y, \|\cdot\|)$ are Banach spaces and $A: X \to Y$ an isomorphism, then Y with the new norm

$$\|y\|_A = \|A^{-1}y\|_X$$

is isometric to X and has the same modulus and characteristic of convexity.

In general it is difficult to describe the modulus of convexity of a Banach space in explicit terms. However in view of (5.5), if E is a two-dimensional subspace of a Banach space X then the following estimates apply: $\delta_X(\epsilon) \leqslant \delta_E(\epsilon)$ and $\epsilon_0(E) \leqslant \epsilon_0(X)$.

We now turn to some specific examples.

Example 5.3 By assigning to $x = (x_1, x_2) \in \mathbb{R}^2$ the respective norms

$$\|x\|_1 = |x_1| + |x_2|;$$
$$\|x\|_\infty = \max\{|x_1|, |x_2|\};$$

one obtains the spaces l_2^1 and l_2^∞. Each of these spaces has a square-shaped unit ball and for each $\delta(\epsilon) = 0$ for $\epsilon \in [0, 2]$. Since any linear isomorphism of $\mathbb{R}^2$ transforms squares into parallelepipeds, any norm generated by a parallelepiped-shaped unit ball has the property $\delta(\epsilon) \equiv 0$ and $\epsilon_0 = 2$. Since the space $\mathscr{C}[0, 1], L^1[0, 1], L^\infty, c$, and c_0 all have two-dimensional subspaces isometric to l_2^1 or l_2^∞ the modulus of convexity of each is identically 0.

The following observation is pertinent to our next example.

Lemma 5.2 *A Banach space X is strictly convex if and only if $\delta(2) = 1$.*

Proof Let $\delta(2) = 1$, and suppose $x, y \in X$ satisfy $\|x\| = \|y\| = \|(x + y)/2\| = 1$. Then

$$\left\|\frac{x - y}{2}\right\| = \left\|\frac{x + (-y)}{2}\right\| \leqslant 1 - \delta(\|x - (-y)\|) = 1 - \delta(2) = 0.$$

Thus $x = y$ and X is strictly convex. On the other hand, suppose X is strictly convex and suppose $\|x\| = \|y\| = 1$ with $\|x - y\| = 2$. Then $x \neq -y$ yields the contradiction: $1 = \|(x - y)/2\| = \|(x + (-y))/2\| < 1$. Hence $x = -y$ and $\delta(2) = 1$.

Example 5.4 Let X be $\mathscr{C}[0, 1]$ with the strictly convex norm $\|\cdot\|_\mu$ of Example 5.1. Thus $\delta_X(2) = 1$. Moreover, each two-dimensional subspace E of X is uniformly convex (since such spaces are strictly convex and finite dimensional). Hence for such E, $\epsilon_0(E) = 0$ and $\delta_E(\epsilon) > 0$ for $\epsilon > 0$. Implicit in the observations of Example 5.1 is the fact that X has two-dimensional subspaces which are 'almost square', so it follows that $\delta_X(\epsilon) = 0$ for $\epsilon \in [0, 2)$, $\delta(2) = 1$, and $\epsilon_0(X) = 2$.

The above example shows that δ_X can actually be discontinuous at 2. In spite of this, we have

$$\delta(2^-) = \lim_{\epsilon \to 2^-} \delta(\epsilon) = 1 - \frac{\epsilon_0}{2}.$$

To see this, let $\epsilon \in [\epsilon_0, 2)$ and take $\eta \in (0, 1 - \delta(\epsilon))$. Select x, y in the unit ball of X satisfying $\|x - y\| = \epsilon$ and

$$\left\|\frac{x+y}{2}\right\| \geq 1 - \delta(\epsilon) - \eta.$$

Then

$$\frac{\epsilon}{2} = \left\|\frac{x-y}{2}\right\| \leq 1 - \delta(\|x - (-y)\|)$$

$$= 1 - \delta(\|x + y\|)$$

$$\leq 1 - \delta(2(1 - \delta(\epsilon) - \eta))$$

and since η is arbitrary,

$$\frac{\epsilon}{2} \leq 1 - \delta(2(1 - \delta(\epsilon))). \tag{5.7}$$

Letting $\epsilon \to 2^-$ we have

$$\delta(2(1 - \delta(2^-))) = 0;$$

hence $\epsilon_0 \geq 2(1 - \delta(2^-))$ from which $\delta(2^-) \geq 1 - (\epsilon_0/2)$. Since the reverse inequality is obtained by letting $\epsilon \to \epsilon_0^+$, we conclude:

$$\delta(2^-) = 1 - \frac{\epsilon_0}{2}.$$

The inequality (5.7) has another interesting consequence. Setting $t = 2(1 - \delta(\epsilon))$, two applications of (5.7) yield

$$1 - \delta(\epsilon) \geq 1 - \delta(2(1 - \delta(t))) \geq \frac{t}{2} = 1 - \delta(\epsilon).$$

Thus equality holds, and since t can assume any value in $[\epsilon_0, 2]$ we conclude that for all $\epsilon \in [\epsilon_0, 2]$:

$$\delta(2(1 - \delta(\epsilon))) = 1 - \frac{\epsilon}{2}. \tag{5.8}$$

Now define $\mu: [\epsilon_0/2, 1] \to [0, 1 - \epsilon_0/2]$ by setting $\mu(\epsilon) = \delta(2\epsilon)$. Then (5.8) implies

$$\mu(\epsilon) + \mu^{-1}(1 - \epsilon) = 1$$

which, in turn, shows that the graph of μ is symmetric about the line $\mu + \epsilon = 1$.

The modulus and coefficient of convexity in some sense scale the convexity of Banach spaces. Roughly speaking, if X and Y are two Banach spaces and if $\delta_X(\epsilon) \geqslant \delta_Y(\epsilon)$ for all $\epsilon \in [0, 2]$ then X is 'more convex' or 'more rotund' than Y. If $\epsilon_0(X) \leqslant \epsilon_0(Y)$ then Y is 'more square' than X.

The following concept (formulated in our terms) is due to R. C. James.

Definition 5.4 A Banach space X is said to be *uniformly nonsquare* if $\epsilon_0(X) < 2$.

This is a significant concept in the theory of Banach spaces. R. C. James (1964b) initially proved that all uniformly nonsquare spaces are reflexive, and later P. Enflo (1972) proved that any such space actually has an equivalent uniformly convex norm. These facts are intimately connected with the following concept (see James, 1972).

Definition 5.5 A Banach space X is said to be *superreflexive* if any Banach space Y which is finitely representable in X is itself reflexive. (Y is *finitely representable* in X if every finite dimensional subspace Y_0 of Y is 'almost isometric' to a subspace of X in the sense that for any $\lambda > 1$ there exists an isomorphism $T: Y_0 \to X$ such that

$$\lambda^{-1} \| y \| \leqslant \| Ty \| \leqslant \lambda \| y \|$$

for all $y \in Y$.)

R. C. James proved that X is superreflexive if and only if it can be given an equivalent uniformly convex norm. This fact combined with Enflo's result yields the following deep result:

Theorem 5.1 (James, Enflo) *For a Banach space X the following are equivalent*:

(a) *X is superreflexive.*
(b) *X has an equivalent uniformly nonsquare norm.*
(c) *X has an equivalent uniformly convex norm.*

There are reflexive Banach spaces which fail to be superreflexive. The most well-known example is the following, due to M. M. Day (1941).

Example 5.5 Let $n \in \mathbb{N}$, and for $x = (x_1, \ldots, x_n) \in \mathbb{R}^n$ set $|x|_n^1 = \Sigma_{i=1}^n |x_i|$ and $|x|_n^\infty = \max\{|x_1|, \ldots, |x_n|\}$. Let $l_n^1 = (\mathbb{R}^n, |\cdot|_n^1)$ and $L_n^\infty = (\mathbb{R}^n, |\cdot|_n^\infty)$, and form the l^2 products of the spaces $\{l_n^1\}$ and $\{L_n^\infty\}$ as follows:

$$D_1 = \left\{ x = \{x^n\}_{n=1}^\infty : x^n \in l_n^1, \left(\sum_{i=1}^\infty (|x^i|_i^1)^2 \right)^{1/2} = \|x\|_{D_1} < \infty \right\};$$

$$D_\infty = \left\{ x = \{x^n\}_{n=1}^\infty : x^n \in L_n^\infty, \left(\sum_{i=1}^\infty (|x^i|_i^\infty)^2 \right)^{1/2} = \|x\|_{D_\infty} < \infty \right\};$$

Each of the spaces D_1 and D_∞ is reflexive and each is the dual of the other. However, $\delta_{D_1}(\epsilon) = \delta_{D_\infty}(\epsilon) = 0$ for all $\epsilon \in [0, 2]$ (i.e., $\epsilon_0(D_1) = \epsilon_0(D_\infty) = 2$), and this remains true under all possible renormings.

We conclude our observations about general properties of the modulus δ with three additional examples.

Example 5.6 Let H denote a Hilbert space with its usual norm. Since H is characterized by the parallelogram law:

$$\|x+y\|^2 + \|x-y\|^2 = 2(\|x\|^2 + \|y\|^2),$$

if $x, y \in H$ satisfy $\|x\| = \|y\| = 1$ and $\|x-y\| = \epsilon$, then

$$\left\| \frac{x+y}{2} \right\| = (1-(\epsilon/2)^2)^{1/2}.$$

Consequently,

$$\delta_H(\epsilon) = 1 - (1-(\epsilon/2)^2)^{1/2}.$$

Now let $\lambda \geqslant 1$ and let X_λ denote the space obtained by renorming the Hilbert space $(l^2, \|\cdot\|)$ as follows. For $x = (x_1, x_2, \ldots) \in l^2$, set

$$\|x\|_\lambda = \max\{\|x\|_\infty, \lambda^{-1}\|x\|\}$$

$$= \max\left\{ \max|x_i|, \lambda^{-1} \left(\sum_{i=1}^\infty x_i^2 \right)^{1/2} \right\}.$$

All the norms $\|\cdot\|_\lambda$ are equivalent since

$$\lambda^{-1}\|x\| \leqslant \|x\|_\lambda \leqslant \|x\|.$$

However for $\lambda > 1$ these norms are not uniformly convex. Routine calculations actually show:

$$\epsilon_0(X_\lambda) = \begin{cases} 2(\lambda^2 - 1)^{1/2} & \text{for } \lambda \leqslant 2^{1/2} \\ 2 & \text{for } \lambda \geqslant 2^{1/2}. \end{cases}$$

Thus, for example, $\epsilon_0(X_{5^{1/2}/2}) = 1$.

Example 5.7 Clarkson (1936) proved that all the classical spaces $L^p = L^p(\mu)$ for $1 < p < \infty$ are uniformly convex. Points $x, y \in L^p$ for $p \geqslant 2$ satisfy the following analogue of the parallelogram law:

$$\|x+y\|^p + \|x-y\|^p \leqslant 2^{p-1}(\|x\|^p + \|y\|^p).$$

Consequently (see Hanner, 1956),

$$\delta_{L^p}(\epsilon) \geqslant 1 - (1 - (\epsilon/2)^p)^{1/p}.$$

The situation for $1 < p < 2$ is more complicated. For such p, Clarkson obtained the inequality

$$\|x+y\|^q + \|x-y\|^q \leqslant 2(\|x\|^p + \|y\|^p)^{1/p},$$

$x, y \in L^p(\mu)$, where $p^{-1} + q^{-1} = 1$. Clarkson used this to obtain an estimate for the modulus of convexity. Hanner (1956) later established the following precise implicit formula for $\delta(\epsilon) = \delta_{L^p}(\epsilon)$, $1 < p < 2$:

$$\left(1 - \delta(\epsilon) + \frac{\epsilon}{2}\right)^p + \left|1 - \delta(\epsilon) - \frac{\epsilon}{2}\right|^p = 2.$$

Hanner derived this fact by showing that the following holds in such spaces:

$$\left|\|x\| - \|y\|\right|^p + \left|\|x\| + \|y\|\right|^p \leqslant \|x-y\|^p + \|x+y\|^p, \qquad x, y \in L^p(\mu).$$

(Lim has recently proved (1986) that, in fact, the above Hanner inequality may be derived from the corresponding Clarkson inequality.)

As a final remark we note that, among all Banach spaces, a Hilbert space H is 'most convex' in the sense that for any space X, $\delta_X(\epsilon) \leqslant \delta_H(\epsilon)$ for each $\epsilon \in [0, 2]$. Nordlander first proved this in 1960; another proof may be found in Diestel (1975). (Obviously only the two-dimensional case need be considered.) We note that the relation $\delta_X(\epsilon) \leqslant \delta_H(\epsilon)$ also follows quickly from a result of Dvoretsky (1961) which states that any infinite dimensional Banach space contains, for any n, a subspace which approximates l_n^2, i.e., for any $\lambda > 1$ there exists a subspace X_n of X and an isomorphism $T: X_n \to l_n^2$ satisfying

$$\lambda^{-1} \|x\| \leqslant \|Tx\|_2 \leqslant \lambda \|x\|, \qquad x \in X.$$

Obviously $\delta_X(\epsilon) \leqslant \delta_{X_n}(\epsilon)$ for $\epsilon \in [0, 2]$, and the latter can be made arbitrarily near $\delta_H(\epsilon)$ by choosing λ sufficiently near 1.

Example 5.8 Let $Q_i, i = 1, \ldots, 4$, denote the ith quadrant in $\mathbb{R}^2$, and for

$x=(x_1,x_2)\in\mathbb{R}^2$, set

$$\|x\|=\begin{cases}(x_1^2+x_2^2)^{1/2} & \text{if } x\in Q_1\cup Q_3 \\ |x_1|+|x_2| & \text{if } x\in Q_2\cup Q_4\end{cases}$$

$$=\begin{cases}\|x\|_2 & \text{if } x_1\cdot x_2\geqslant 0 \\ \|x\|_1 & \text{if } x_1\cdot x_2\leqslant 0\end{cases}$$

$$=\max\{x_1-x_2,x_2-x_1,(x_1^2+x_2^2)^{1/2}\}.$$

Thus the unit ball B of $(\mathbb{R}^2,\|\cdot\|)$ is contained in the ordinary unit disc and for any $x\in\mathbb{R}^2$,

$$\|x\|_2\leqslant\|x\|\leqslant 2^{1/2}\|x\|_2. \tag{5.9}$$

We now evaluate the modulus of convexity δ for $(\mathbb{R}^2,\|\cdot\|)$. Note first that $\epsilon_0=2^{1/2}$. We consider two cases:

Case 1 Take $\epsilon>2^{1/2}$ and $\|x\|=\|y\|=1$ with $\|x-y\|\geqslant\epsilon$, where $y-x\in Q_1\cup Q_3$. Thus $\|x\|_2\leqslant 1$, $\|y\|_2\leqslant 1$, and $\|x-y\|_2\geqslant\epsilon$, which implies $\frac{1}{2}\|x+y\|_2\leqslant[1-(\epsilon/2)^2]^{1/2}$. This in turn implies $\frac{1}{2}\|x+y\|\leqslant 2^{-1/2}\|x+y\|_2\leqslant(2-\epsilon^2/2)^{1/2}$; thus $1-\frac{1}{2}\|x+y\|\geqslant 1-(2-\epsilon^2/2)^{1/2}$.

Case 2 Now take ϵ,x,y as above, but with $y-x\in Q_2\cup Q_4$. Observe that now $\frac{1}{2}(x+y)\in Q_1\cup Q_3$. Since $\|x-y\|\geqslant\epsilon$ implies $\|x-y\|_2\geqslant\epsilon/2^{1/2}$, we have

$$\frac{1}{2}\|x+y\|=\frac{1}{2}\|x+y\|_2\leqslant(1-\epsilon^2/8)^{1/2};$$

thus

$$1-\frac{1}{2}\|x+y\|\geqslant 1-(1-\epsilon^2/8)^{1/2}$$

It follows that for $\epsilon>2^{1/2}$,

$$\delta(\epsilon)\geqslant\max\{1-(2-\epsilon^2/2)^{1/2},1-(1-\epsilon^2/8)^{1/2}\}. \tag{5.10}$$

Considering points x,y such that $y-x$ has the direction of one of the diagonals ($\pi/4$ or $-\pi/4$) reveals that actually equality occurs in (5.10). Thus the modulus of convexity for $(\mathbb{R},\|\cdot\|)$ is the *nonconvex* function:

$$\delta(\epsilon)=\begin{cases}0 & \text{if } 0\leqslant\epsilon\leqslant 2^{1/2} \\ \max\{1-(2-\epsilon^2/2)^{1/2},1-(1-\epsilon^2/8)^{1/2}\} & \text{if } 2^{1/2}\leqslant\epsilon\leqslant 2.\end{cases}$$

6

The modulus of convexity and normal structure

There is one fundamental implication connecting the modulus of convexity of a Banach space X with normal structure.

Theorem 6.1 *If the modulus of convexity δ of a Banach space X satisfies $\delta(1) > 0$ (i.e. if $\epsilon_0(X) < 1$), then X has normal structure.*

Proof Let $K \subset X$ be closed and convex with diam $K = d > 0$. Take $\mu > 0$ so that $d - \mu > \epsilon_0(X)d$, select $u, v \in K$ such that $\|u - v\| \geqslant d - \mu$, and set $z = \frac{1}{2}(u + v)$. Then for any $x \in K$: $\|x - u\| \leqslant d$, $\|x - v\| \leqslant d$, while $\|u - v\| \geqslant d - \mu$. Consequently

$$\|x - z\| \leqslant \left[1 - \delta\left(\frac{d - \mu}{d}\right) \right] d \tag{6.1}$$

and, since $\delta[(d - \mu)/d] > 0$, z is a nondiametral point of K.

Implicit in the above proof is the fact that if X is uniformly convex $(\epsilon_0(X) = 0)$ then for any $u, v \in K$ with $u \neq v$ the point $z = \frac{1}{2}(u + v)$ is a nondiametral point of K. Therefore in uniformly convex spaces all diametral points of K are 'extremal points' of K. (A point of K is *extremal* if it must be one of the endpoints of any segment lying in K which contains it.)

It is actually possible to draw stronger conclusions from the proof of Theorem 6.1 than the one stated. These conclusions are related to the notion of uniform normal structure, discussed in Chapter 4, and to the following concept.

Definition 6.1 The *normal structure coefficient* (Bynum, 1980) of a Banach space X is the number:

$$N(X) = \sup\left\{ \frac{r(K)}{\text{diam}\,(K)} : K \subset X \text{ is bounded and convex, diam } K > 0 \right\}.$$

In other words, $N(X)$ is the smallest number for which $r(K) \leqslant N(X)$ diam K for all bounded convex subsets K of X.

Obviously $N(X) \leqslant 1$, while $N(X) < 1$ if and only if X has uniform normal structure.

We now observe that (6.1) implies the following stronger version of Theorem 6.1.

Theorem 6.1′ *If the modulus of convexity* δ *of a Banach space satisfies* $\delta(1) > 0$, *then* X *has uniform normal structure and*

$$N(X) \leqslant 1 - \delta(1).$$

In particular, it is always the case that

$$r(K) \leqslant (1 - \delta(1)) \operatorname{diam} K, \tag{6.2}$$

and this fact yields an estimate for the diameter of the Chebyshev center $C(K)$ of K. To see this take $\mu > 0$ and select $u, v \in C(K)$ such that

$$\|u - v\| \geqslant (1 - \mu) \operatorname{diam} C(K).$$

Now set $z = \frac{1}{2}(u + v)$ and select $x \in K$ such that $\|z - x\| \geqslant (1 - \mu) r(K)$. Then, $\|x - u\| \leqslant r(K)$ and $\|x - v\| \leqslant r(K)$, and thus

$$(1 - \mu) r(K) \leqslant \|z - x\| \leqslant \left(1 - \delta \left(\frac{(1 - \mu) \operatorname{diam} C(K)}{r(K)} \right) \right) r(K).$$

This implies (letting $\mu \to 0^+$) $\delta[\operatorname{diam} C(K) / r(K)] \leqslant 0$, i.e.,

$$\operatorname{diam} C(K) \leqslant \epsilon_0(X) r(K).$$

In view of (6.2)

$$\operatorname{diam} C(K) \leqslant \epsilon_0(X)(1 - \delta(1)) \operatorname{diam} K. \tag{6.3}$$

The above specifically shows that the Chebyshev center of any bounded convex subset of a uniformly convex space always consists of exactly one point.

The estimate in (6.2) is, in general, not sharp.

Example 6.1 For a Hilbert space H, $1 - \delta(\epsilon) = (1 - \epsilon^2/4)^{1/2}$, so (6.2) implies $N(H) \leqslant 3^{1/2}/4$. However, it is possible to show that $r(K) \leqslant (2^{1/2}/2) \operatorname{diam} K$ and that this inequality is the best possible. Indeed, if K is the positive part of the unit ball in l^2, then

$$K = \{ x = \{ x_i \} : \|x\| \leqslant 1, x_i \geqslant 0, i = 1, 2, \dots \};$$

thus $r(K) = 1$ while $\operatorname{diam} K = 2^{1/2}$.

A general formula for $N(X)$ in arbitrary Banach spaces is not known; this coefficient is fully understood only in certain special cases.

In view of Theorem 6.1 it is natural to consider whether $\epsilon_0(X) = 1$ implies normal structure. The following example, due to Bynum (1972), shows that the condition $\epsilon_0(X) < 1$ is sharp.

Example 6.2 Consider the classical $l^p = l^p(\mathbb{N})$ spaces for $1 \leqslant p < +\infty$, consisting of sequences $x = \{x_n\}$ satisfying

$$\|x\|_p = \left(\sum_{n=1}^{\infty} |x_n|^p \right)^{1/p} < +\infty.$$

Each element $x \in l^p$ may be represented as $x = x^+ - x^-$ where the respective ith components of x^+ and x^- are given by

$$(x^+)_i = \max\{x_i, 0\} = \frac{x_i + |x_i|}{2};$$

$$(x^-)_i = \max\{0, -x_i\} = \frac{|x_i| - x_i}{2}.$$

Now let $p, q \in [1, +\infty)$ be arbitrary, and for $x \in l^p$ set

$$\|x\|_{p,q} = (\|x^+\|_p^q + \|x^-\|_p^q)^{1/q};$$

$$\|x\|_{p,\infty} = \max\{\|x^+\|_p, \|x^-\|_p\}.$$

The above procedure provides l^p with a family of equivalent norms, and the spaces so generated will be denoted by $l^{p,q}$. The spaces $l^{p,q}$ are uniformly convex if $p, q \in (1, \infty)$, while $\epsilon_0(l^{p,\infty}) = 1$ and $\epsilon_0(l^{p,1}) = 2^{1/p}$. The space $l^{p,\infty}$ fails to have normal structure, showing that the assumption $\epsilon_0(X) < 1$ in Theorem 6.1 is sharp. On the other hand, the space $l^{p,1}$ has normal structure in spite of the fact that $\epsilon_0(l^{p,1}) > 1$. Moreover, the spaces $l^{p,\infty}$ and $l^{p,1}$ are dual to each other. This shows that *normal structure is not a condition which is invariant under passing to dual spaces*.

Our next observation is that uniform normal structure is stable under small norm perturbations.

Theorem 6.2 *Let X be a Banach space and let $X_1 = (X, \|\cdot\|_1)$ and $X_2 = (X, \|\cdot\|_2)$, where $\|\cdot\|_1$ and $\|\cdot\|_2$ are two equivalent norms on X satisfying for $\alpha, \beta > 0$,*

$$\alpha\|x\|_1 \leqslant \|x\|_2 \leqslant \beta\|x\|_1, \qquad x \in X.$$

If $k = \beta/\alpha$, then

$$(1/k)N(X_1) \leqslant N(X_2) \leqslant kN(X_1). \tag{6.4}$$

Proof This is a straightforward consequence of the definitions.

Under the assumptions of Theorem 6.2 it is possible to determine certain relations between the moduli of convexity $\delta_1 = \delta_{X_1}$ and $\delta_2 = \delta_{X_2}$. We note the following implications for $x, y \in X$ and $\epsilon > 0$:

$$\left.\begin{array}{c} \|x\|_2 \leqslant 1 \\ \|y\|_2 \leqslant 1 \\ \|x-y\|_2 \geqslant \epsilon \end{array}\right\} \Rightarrow \left.\begin{array}{c} \|x\|_1 \leqslant \alpha^{-1} \\ \|y\|_1 \leqslant \alpha^{-1} \\ \|x-y\|_1 \geqslant \epsilon\beta^{-1} \end{array}\right\} \Rightarrow$$

$$\left\|\frac{x+y}{2}\right\|_1 \leqslant (1 - \delta_1(\epsilon k^{-1}))\alpha^{-1} \Rightarrow \left\|\frac{x+y}{2}\right\|_2 \leqslant k(1 - \delta_1(\epsilon k^{-1})).$$

From this it follows that

$$\delta_2(\epsilon) \geqslant 1 - k(1 - \delta_1(\epsilon k^{-1})).$$

Consequently,

Theorem 6.3 *Under the assumptions of Theorem 6.2, if $\epsilon_0(X_1) < 1$ and if*

$$k(1 - \delta_1(k^{-1})) < 1, \tag{6.5}$$

then $\epsilon_0(X_2) < 1$.

Our next example shows that the above observations are sharp.

Example 6.3 Consider $X_1 = l^2$ and the spaces $X_\lambda, \lambda > 1$, as defined in Example 5.6. Then $k = k_\lambda = \lambda$ and, since $\delta_{l^2}(\epsilon) = 1 - (1 - (\epsilon/2)^2)^{1/2}$, (6.5) implies $\lambda < 5^{1/2}/2$. Actually, $\epsilon_0(X_\lambda)$ increases from 0 to 2 as λ increases from 1 to $2^{1/2}$. Also, for $5^{1/2}/2 \leqslant \lambda < 2^{1/2}$, Theorem 6.2 implies $N(X_\lambda) \leqslant kN(X_1) = \lambda N(X_1) < 1$, so X_λ has uniform normal structure for $1 < \lambda < 2^{1/2}$. Yet for $5^{1/2}/2 < \lambda < 2^{1/2}$,

$$1 < \epsilon_0(X_\lambda) = 2(\lambda^2 - 1)^{1/2} < 2.$$

For $\lambda = 2^{1/2}$, $\epsilon_0(X_\lambda) = 2$ and $k = 2^{1/2}$. Here X_λ fails to have normal structure since the set $K \subset X_{2^{1/2}}$ defined by

$$K = \left\{ x = \{x_i\} : x_i \geqslant 0, \sum_{i=1}^{\infty} x_i^2 \leqslant 1 \right\}$$

satisfies $r(K) = \operatorname{diam} K = 1$ with respect to the $X_{2^{1/2}}$ norm.

The above example shows that, in particular, a uniformly convex space (Hilbert space) has an equivalent norm which fails to have normal structure.

On the other hand, as we have previously noted, every superreflexive space as an equivalent uniformly convex norm. This fact allows us to show that in a sense, it is very unusual for spaces to fail to have normal structure. We begin with the following:

Lemma 6.1 *Suppose* $(X, \|\cdot\|)$ *is a superreflexive space and suppose* $\|\cdot\|_0$ *is a uniformly convex norm on* X *which is equivalent to* $\|\cdot\|$*. Then for any* $\mu > 0$*, the norm* $\|\cdot\|_\mu$ *defined by*

$$\|x\|_\mu = \|x\| + \mu\|x\|_0, \qquad x \in X,$$

is also uniformly convex.

Proof It suffices to show that for $\{x_n\}$ and $\{y_n\}$ in X, the conditions

$$\lim_{n \to \infty} \|x_n\|_\mu = \lim_{n \to \infty} \|y_n\|_\mu = \lim_{n \to \infty} \tfrac{1}{2}\|x_n + y_n\|_\mu = 1 \tag{6.6}$$

imply $\lim_{n \to \infty} \|x_n - y_n\|_\mu = 0$. To see this, note that (6.6) implies

$$
\begin{aligned}
1 &= \lim_{n \to \infty} \tfrac{1}{2}\|x_n + y_n\|_\mu \\
&= \tfrac{1}{2} \liminf_{n \to \infty} (\|x_n + y_n\| + \mu\|x_n + y_n\|_0) \\
&\leqslant \tfrac{1}{2} \liminf_{n \to \infty} (\|x_n\| + \|y_n\| + \mu\|x_n + y_n\|_0) \\
&= \tfrac{1}{2} \liminf_{n \to \infty} \{\|x_n\|_\mu + \|y_n\|_\mu + \mu[\|x_n + y_n\|_0 - (\|x_n\|_0 + \|y_n\|_0)]\} \\
&= 1 - \limsup_{n - \infty} \mu(\|x_n\|_0 + \|y_n\|_0 - \|x_n + y_n\|_0).
\end{aligned}
$$

Thus $\limsup_{n \to \infty}(\|x_n\|_0 + \|y_n\|_0 - \|x_n + y_n\|_0) = 0$ and, since $\|\cdot\|_0$ is uniformly convex, it follows that $\lim_{n \to \infty} \|x_n - y_n\|_0 = 0$. Hence (since $\|\cdot\|_0$ is equivalent to $\|\cdot\|_\mu$), $\lim_{n \to \infty} \|x_n - y_n\|_\mu = 0$.

As a consequence of the above, any norm on X can be closely approximated by a uniformly convex norm. We formalize this as follows:

For two equivalent norms $\|\cdot\|_1, \|\cdot\|_2$ on a Banach space X, set

$$\mathcal{K}(\|\cdot\|_1, \|\cdot\|_2) = \inf\{\beta/\alpha \colon \alpha\|x\|_1 \leqslant \|x\|_2 \leqslant \beta\|x\|_1 \text{ for all } x \in X\}.$$

In other words,

$$
\begin{aligned}
\mathcal{K}(\|\cdot\|_1, \|\cdot\|_2) &= \sup_{x \neq 0}(\|x\|_2/\|x\|_1) \sup_{x \neq 0}(\|x\|_1/\|x\|_2) \\
&= \|I_1\| \cdot \|I_2\|,
\end{aligned}
$$

where the last two numbers denote the operator norms of the identity viewed respectively as a mapping from $(X, \|\cdot\|_1)$ to $(X, \|\cdot\|_2)$ and from $(X, \|\cdot\|_2)$ to $(X, \|\cdot\|_1)$. The function $\mathcal{K}$ is called the Banach–Mazur distance between equivalent norms; this distance is not a metric since $\mathcal{K}(\|\cdot\|_1, \|\cdot\|_1) = 1$. However, it is symmetric and, since

$$\mathcal{K}(\|\cdot\|_1, \|\cdot\|_3) \leqslant \mathcal{K}(\|\cdot\|_1, \|\cdot\|_2) \cdot \mathcal{K}(\|\cdot\|_2, \|\cdot\|_3),$$

the family $\mathcal{N}$ of equivalent norms on X may be metrized by taking

$$\rho(\|\cdot\|_1, \|\cdot\|_2) = \ln \mathcal{K}(\|\cdot\|_1, \|\cdot\|_2).$$

In this context, our earlier observations may be summarized as follows:

Theorem 6.4 *If $(X, \|\cdot\|)$ is a superreflexive Banach space and if $\mathcal{N} = (\mathcal{N}, \rho)$ denotes the space of all norms on X equivalent to $\|\cdot\|$, then*

(a) *The family of all uniformly convex norms in $\mathcal{N}$ is dense in $\mathcal{N}$;*
(b) *The family of all norms in $\mathcal{N}$ for which $\epsilon_0 < 1$ is open and dense in $\mathcal{N}$; and*
(c) *The family of all norms in $\mathcal{N}$ with respect to which X has uniform normal structure is open and dense in $\mathcal{N}$.*

Spaces which are not superreflexive do not have equivalent uniformly convex norms. However the above ideas can be extended further using a result of V. Zizler. To do this another definition is needed:

For a Banach space $(X, \|\cdot\|)$ and a fixed element $z \in X$ with $\|z\| = 1$, let the *modulus of convexity of X in the direction $z \in X$* be the function $\delta_z : [0, 2] \to [0, 1]$ defined by

$$\delta_z(\epsilon) = \inf\{1 - \tfrac{1}{2}\|x + y\| : \|x\| \leqslant 1, \|y\| \leqslant 1, x - y = \epsilon z\}.$$

If $\delta_z(\epsilon) > 0$ for all $\epsilon > 0$ and all z then X is said to be *uniformly convex in every direction.*

Zizler has shown that a space X may be uniformly convex in every direction while failing to be uniformly convex. (It may even happen that $\epsilon_0(X) = 2$.) Obviously, such spaces are always strictly convex.

Recall that a family $\mathcal{J}$ of functionals on a space is said to *separate points* of X if for each $x, y \in X$ with $x \neq y$ there exists $f \in \mathcal{J}$ such that $f(x) \neq f(y)$.

Theorem 6.5 (Zizler, 1971) *Suppose $(X, \|\cdot\|)$ is a Banach space whose dual space X^* contains a countable family $\mathcal{J}$ which separates points of X. Then there exists a norm $\|\cdot\|_0$ on X which is equivalent to $\|\cdot\|$ and which is uniformly convex in every direction.*

Zizler's construction is as follows: Let $S = \{f_1, f_2, \dots\}$ and $\mu > 0$. For $x \in X$, set

$$\|x\|_\mu = \left(\|x\|^2 + \mu \sum_{i=1}^\infty \left(\frac{f_i(x)}{2^i} \right)^2 \right)^{1/2}.$$

The above norm is uniformly convex in every direction and for small μ it is close to $\|\cdot\|$ in the metric ρ.

Clearly Zizler's Theorem applies to all separable spaces. (Indeed, if $\{x_1, x_2, \ldots\}$ is a countable dense subset of X, then by the Hahn–Banach Theorem there exist functionals $f_{ij} \in X^*$ such that $f_{ij}(x_i - x_j) = \|x_i - x_j\|$. Obviously the family $\mathcal{J} = \{f_{ij}\}$ separates points of X.)

We may now formulate the following.

Theorem 6.5 *Suppose X satisfies the assumptions of Zizler's Theorem. Then the family of all norms in $\mathcal{N}$ having normal structure is dense in $\mathcal{N}$.*

Proof Since the family of all norms in $\mathcal{N}$ which are uniformly convex in every direction is dense in $\mathcal{N}$, it suffices to show that all such norms have normal structure. Let K be bounded, closed and convex and suppose $u, v \in K$ with $u \neq v$. Set $z = (u - v)/\|u - v\|$ and $w = \frac{1}{2}(u + v)$. Then for any $x \in K$,

$$\left. \begin{array}{l} \|x - u\| \leqslant \operatorname{diam} K \\ \|x - v\| \leqslant \operatorname{diam} K \end{array} \right\} \Rightarrow \|x - w\| \leqslant \left(1 - \delta_z \left(\frac{\|u - v\|}{\operatorname{diam} K} \right) \right) \operatorname{diam} K,$$

showing that w is not diametral.

Since all separable spaces fall into Zizler's category and since many such spaces are not reflexive, the above theorem shows that nonreflexive spaces may have normal structure even if they have characteristic of convexity $\epsilon_0 = 2$ (see also Day, James and Swaminathan, 1971).

We conclude with:

Example 6.4 Consider $\mathscr{C}[0, 1]$ and let $\{t_n\}$ be a sequence which is dense in $[0, 1]$. The norm $\|\cdot\|_\mu$ for $\mu > 0$ defined on $\mathscr{C}[0, 1]$ by setting

$$\|x\|_\mu = \max|x(t)| + \mu \left[\sum_{i=1}^\infty \left(\frac{x(t_i)}{2^i} \right)^2 \right]^{1/2}, \qquad x \in \mathscr{C}[0, 1],$$

is uniformly convex in every direction. Since $\mathscr{C}[0, 1]$ is universal with respect to all separable Banach spaces, this also shows that such renormings exist for any separable space via an isometric embedding in $\mathscr{C}[0, 1]$. Also, returning to Example 2.1, we see that the mapping $T: K \to K$ defined on

$$K = \{x \in \mathscr{C}[0, 1] : 0 = x(0) \leqslant x(t) \leqslant x(1) = 1\}$$

by taking $(Tx)(t) = tx(t)$ is not only nonexpansive but also contractive with respect to $\|\cdot\|_\mu$. This example shows that the fixed point property for nonexpansive mappings does not follow from normal structure alone. Either weak compactness or some other assumption is essential.

7

Normal structure and smoothness

As we have seen, normal structure is not a dual invariant. And while in Bynum's example (Example 6.2), both the space X and its dual X^* have coefficient $\epsilon_0 \geqslant 1$, even this fails to be the case in general.

We begin with some standard definitions.

Definition 7.1 A Banach space X is said to be *smooth* if for every $x \in X$ with $\|x\| = 1$, there exists a unique $x^* \in X^*$ such that $\|x^*\| = x^*(x) = 1$.

It is not difficult to prove that a Banach space X is smooth if and only if for every $x, y \in X$ with $x \neq 0$ the following limit exists:

$$\lim_{t \to 0} t^{-1}[\|x + ty\| - \|x\|] = \varphi_x(y). \tag{7.1}$$

This limit defines a functional $\varphi_x \in X^*$ which is called the *Gateaux derivative* of the norm at x.

Definition 7.2 A Banach space X is called *uniformly smooth* if the limit (7.1) exists uniformly in the set $\{(x, y): \|x\| = \|y\| = 1\}$; thus X is uniformly smooth if for each $\epsilon > 0$ there exists $\delta > 0$ such that for $|t| < \delta$ and all $x, y \in X$ with $\|x\| = \|y\| = 1$,

$$\big|\|x + ty\| - \|x\| - \varphi_x(y)\big| < \epsilon|t|.$$

If the limit (7.1) exists uniformly for $\|y\| = 1$ when x is fixed, then the norm of X is said to be *Fréchet differentiable*.

As with the modulus of convexity, it is possible to scale Banach spaces with respect to their 'smoothness' by introducing the notion of modulus of smoothness.

Definition 7.3 The *modulus of smoothness* of a Banach space X is the function $\rho_X \colon [0, \infty) \to [0, \infty)$ defined by

$$\rho_X(\tau) = \sup\{\tfrac{1}{2}[\|x + \tau y\| + \|x - \tau y\|] - 1 : \|x\| = \|y\| = 1\}. \tag{7.2}$$

69

It is easy to see that a space X is uniformly smooth if and only if

$$\rho'_X(0) = \lim_{\tau \to 0} \frac{\rho_X(\tau)}{\tau} = 0. \tag{7.3}$$

This limit plays a role in scaling 'smoothness' similar to that of $\epsilon_0(X)$ in scaling convexity.

Theorem 7.1 (Lindenstrauss and Tzafriri, 1977, Vol. II) *For any Banach space* X:

(a) $\rho_{X^*}(\tau) = \sup\{(\tau\epsilon/2) - \delta_X(\epsilon): 0 \leqslant \epsilon \leqslant 2\}$ *for all* $\tau > 0$.
(b) $\rho'_{X^*}(0) = \lim_{\tau \to 0}(\rho_{X^*}(\tau)/\tau) = (\epsilon_0(X)/2)$.
(c) X *is uniformly convex if and only if* X^* *is uniformly smooth.*

Proof For $\tau > 0$, $x, y \in X$, and $x^*, y^* \in X^*$,

$$
\begin{aligned}
2\rho_{X^*}(\tau) &= \sup\{\|x^* + \tau y^*\| + \|x^* - \tau y^*\| - 2: \|x^*\| = \|y^*\| = 1\} \\
&= \sup\{x^*(x) + \tau y^*(x) + x^*(y) - \tau y^*(y) \\
&\qquad - 2: \|x\| = \|y\| = \|x^*\| = \|y^*\| = 1\} \\
&= \sup\{\|x + y\| + \tau\|x - y\| - 2: \|x\| = \|y\| = 1\} \\
&= \sup\{\|x + y\| + \tau\epsilon - 2: \|x\| = \|y\| = 1; \|x - y\| = \epsilon, 0 \leqslant \epsilon \leqslant 2\} \\
&= \sup\{\tau\epsilon - 2\delta_X(\epsilon): 0 \leqslant \epsilon \leqslant 2\}.
\end{aligned}
$$

Assertions (a) and (b) follow easily from the above.

Since all uniformly convex spaces are superreflexive, so also are all uniformly smooth spaces. (Actually, any superreflexive space has an equivalent norm which is both uniformly convex and uniformly smooth (Turett, 1982).)

The question of whether uniformly smooth spaces have normal structure remained open for some time, even after it was already known (Baillon 1978–79) that the nonempty, bounded, closed and convex subsets of such spaces (and even spaces X for which $\rho'_X(0) < \frac{1}{2}$) had the fixed point property (f.p.p.) for nonexpansive mappings. B. Turett (1982) first observed that such spaces have normal structure; the simple proof we give below is due to S. Prus.

Theorem 7.2 *If a Banach space* X *has the property* $\rho'_X(0) < \frac{1}{2}$, *then* X *is superreflexive and has normal structure.*

Proof Superreflexivity follows from the fact that such spaces actually are dual to spaces for which $\epsilon_0 < 1$; see Turett (1982). This fact is also a consequence of the following observation of R. C. James (1972): If X is not superreflexive then for any $c < 1$ there exist x_1, x_2 in the unit ball of X and x_1^*, x_2^* in the unit ball of X^* such that

$$x_1^*(x_1) = x_1^*(x_2) = x_2^*(x_2) = c; \qquad x_2^*(x_1) = 0.$$

Hence for all $\tau > 0$:

$$\begin{aligned}
\rho_X(\tau) &\geq \tfrac{1}{2}(\|x_2 + \tau x_1\| + \|x_2 - \tau x_1\|) - 1 \\
&\geq \tfrac{1}{2}[x_1^*(x_2 + \tau x_1) + x_2^*(x_2 - \tau x_1)] - 1 \\
&= c(1 + (\tau/2)) - 1.
\end{aligned}$$

Since $c < 1$ is arbitrary, $\rho_X(\tau) \geq \tau/2$.

Now suppose that X does not have normal structure. We observed earlier in discussing diametral sequences that this implies the existence of a sequence $\{x_n\}$ in the unit ball of X for which $w\text{-}\lim_{n \to \infty} x_n = 0$, $\lim_{n \to \infty} \|x_n\| = 1$, while $\operatorname{diam}\{x_1, x_2, \ldots\} \leq 1$. Consider the sequence $\{x_n^*\}$ of norm one functionals for which

$$x_n^*(x_n) = \|x_n\|, \qquad n = 1, 2, \ldots.$$

Since X^* is reflexive we may assume $\{x_n^*\}$ converges weakly to some $x^* \in X^*$. Select i so that $|x^*(x_i)| < \epsilon/2$ while $\|x_n\| > 1 - \epsilon$ for all $n \geq i$. Then for $j > i$ sufficiently large,

$$|(x_j^* - x^*)(x_i)| < \epsilon/2 \qquad \text{and} \qquad |x_i^*(x_j)| < \epsilon.$$

Consequently, $|x_j^*(x_i)| < \epsilon$, and we have for all $\tau \in (0, 1)$:

$$\begin{aligned}
\rho_X(\tau) &\geq \tfrac{1}{2}(\|x_i - x_j + \tau x_i\| + \|x_i - x_j - \tau x_i\|) - 1 \\
&\geq \tfrac{1}{2}(|x_i^*((1 + \tau)x_i - x_j)| + |x_j^*(x_j - (1 - \tau)x_i)|) - 1 \\
&\geq \tfrac{1}{2}((1 + \tau)(1 - \epsilon) - \epsilon + 1 - \epsilon - (1 - \tau)\epsilon) - 1 \\
&= (\tau/2) - 2\epsilon.
\end{aligned}$$

Since $\epsilon > 0$ is arbitrary, $\rho_X(\tau) \geq \tau/2$ from which $\rho_X'(0) \geq \tfrac{1}{2}$, contradicting the hypothesis $\rho_X'(0) < \tfrac{1}{2}$.

Actually, the condition $\rho_X'(0) < \tfrac{1}{2}$ implies uniform normal structure. However, the only known proofs of this fact are due to Prus (1988) and Khamsi (1987) who use nonstandard methods, and so we defer the proof of this fact to Chapter 14.

8

Conditions involving compactness

As it relates to normal structure, the scaling of Banach spaces with respect to the characteristic ϵ_0 is not precise, since the fact that no 'long segments' in the unit ball are close to the unit sphere is not essential for the presence of normal structure. Indeed, a finite dimensional Banach space always has normal structure even if its unit sphere has large 'flat spots'. In infinite dimensional spaces, normal structure is a result of the fact that these flat spots are 'nearly compact'. This is a concept which may be measured in several ways. We begin by considering sequences.

For a bounded sequence $\{x_n\}$ in a Banach space X, define the constant $\mathrm{sep}\{x_n\}$ as follows:

$$\mathrm{sep}\{x_n\} = \inf\{\|x_n - x_m\| : n \neq m\}.$$

Obviously if $\mathrm{sep}\{x_n\} > 0$ then $\{x_n\}$ is not compact and, in particular, does not contain a convergent subsequence. The fact that $\{x_n\}$ may be relatively weakly compact leads to the following:

Definition 8.1 A Banach space X is said to have *Kadec–Klee (KK) norm* if for every sequence $\{x_n\}$ in X the following implication holds:

$$\left. \begin{array}{l} \|x_n\| \leqslant 1 \\[2mm] \mathrm{sep}\{x_n\} > 0 \\[2mm] w\text{-}\lim_{n \to \infty} x_n = x \end{array} \right\} \Rightarrow \|x\| < 1.$$

A space X is said to have *uniformly Kadec–Klee* (UKK) norm if for any $\epsilon > 0$ there exists $\delta > 0$ such that

$$\left. \begin{array}{l} \|x_n\| \leqslant 1 \\[2mm] \mathrm{sep}\{x_n\} > \epsilon \\[2mm] w\text{-}\lim_{n \to \infty} x_n = x \end{array} \right\} \Rightarrow \|x\| \leqslant 1 - \delta.$$

Clearly the second condition implies the first and, while these concepts were introduced for other purposes, reflexive spaces with (UKK) norm may be easily shown to have normal structure. Specifically, such a space cannot contain a weakly convergent diametral sequence. On the other hand, it is worth noting that there are nonreflexive spaces which have (UKK) norm. For example, the (UKK) condition is vacuously satisfied in l^1 because any weakly convergent sequence $\{x_n\}$ in l^1 must converge in norm; thus $\text{sep}\{x_n\} = 0$.

The following definition is due to R. Huff.

Definition 8.2 A Banach space X is said to be *nearly uniformly convex* (NUC) if for any $\epsilon > 0$ there exists $\delta > 0$ such that for any sequence $\{x_n\}$ in X:

$$\left.\begin{array}{l} \|x_n\| \leq 1 \\ \text{sep}\{x_n\} > \epsilon \end{array}\right\} \Rightarrow \text{dist}\{0, \text{conv}\{x_n\}\} < 1 - \delta.$$

It is known (Huff, 1980) that if a space X is (NUC) then it is reflexive; moreover, within the category of reflexive spaces: X is (NUC)$\Leftrightarrow X$ has (UKK) norm. Thus spaces which are (NUC) also have normal structure.

Finally, we note that van Dulst (1984) has observed that normal structure is implied by a condition weaker than (UKK). The statement 'for every $\epsilon > 0$' can be replaced with 'there exists $\epsilon \in (0, 1)$'. Spaces satisfying this weaker property are said to have a *weakly uniformly Kadec–Klee* (WUKK) norm. (See also van Dulst and Sims, 1983.)

An analogous, and also widely studied, classification of Banach spaces due to K. Goebel and T. Sekowski (1984) utilizes the concept of 'measure of noncompactness', a concept first introduced by Kuratowski (1930) as follows: Let (M, ρ) denote a complete metric space and let $\mathscr{B}$ denote the collection of nonempty, bounded subsets of M. Define the *measure of noncompactness* $\alpha: \mathscr{B} \to \mathbb{R}^+$ by taking for $A \in \mathscr{B}$,

$$\alpha(A) = \inf\{\epsilon > 0: A \text{ is contained in the union of a finite}$$
$$\text{number of sets in } \mathscr{B} \text{ each having}$$
$$\text{diameter less than } \epsilon\}.$$

The function α has the following properties, each of which is easy to verify. For $A, B \in \mathscr{B}$:

(a) $\alpha(A) = 0 \Leftrightarrow \bar{A}$ is compact;
(b) $\alpha(A) = \alpha(\bar{A})$;
(c) $A \subset B \Rightarrow \alpha(A) \leq \alpha(B)$;
(d) $\alpha(A \cup B) = \max\{\alpha(A), \alpha(B)\}$;
(e) $\alpha(A \cap B) \leq \min\{\alpha(A), \alpha(B)\}$;

and (very important)

(f)   if $\{A_n\}$ is a descending sequence $(A_{n+1} \subset A_n)$ of closed, nonempty subsets of $\mathscr{B}$ and if $\lim_{n \to \infty} \alpha(A_n) = 0$, then $A_\infty = \bigcap_{n=1}^\infty A_n$ is nonempty and compact.

Three more properties hold if M is a Banach space, the case in which we are most interested:

(g) $\alpha(A + B) \leqslant \alpha(A) + \alpha(B)$;

(h) $\alpha(cA) = |c|\alpha(A), c \in \mathbb{R}$;

and (also very important);

(i) $\alpha(\text{conv } A) = \alpha(A)$.

There are other ways to define real-valued functions satisfying properties (a)–(i). One of the most commonly used is the so-called *Hausdorff* (or *ball*) *measure of noncompactness*. This is the function $\chi: \mathscr{B} \to \mathbb{R}^+$ defined for $A \in \mathscr{B}$ by:

$$\chi(A) = \inf\{r > 0: A \text{ can be covered by a finite number}$$
$$\text{of balls centered in } M \text{ of radius } r\}.$$

The name Hausdorff is associated with this measure for the following reason. Let $(\mathscr{M}, D)$ denote the (complete) metric space consisting of all closed subsets of M with the Hausdorff metric D and, as before, let $\mathscr{N}$ denote the family of compact subsets of M. Then for $A \in \mathscr{B}, \chi(A) = \text{dist}\{\bar{A}, \mathscr{N}\}$ in $\mathscr{M}$.

We note that the two measures α and χ are equivalent in the following sense:

$$\chi(A) \leqslant \alpha(A) \leqslant 2\chi(A).$$

An even sharper relation holds in some spaces; for example, in Hilbert space

$$2^{1/2}\chi(A) \leqslant \alpha(A) \leqslant 2\chi(A).$$

It is easy to prove that for any ball $B(x; r)$ in an infinite dimensional Banach space, $\chi(B(x; r)) = r$. A more technical argument yields: $\alpha(B(x; r)) = 2r$. Moreover, for any $A \in \mathscr{B}$ the following implications hold: If A contains a sequence $\{x_n\}$ with $\text{sep}\{x_n\} > \epsilon$ then $\alpha(A) > \epsilon$, while $\chi(A) > \epsilon$ implies A contains a sequence $\{x_n\}$ with $\text{sep}\{x_n\} > \epsilon$.

In general it is difficult to determine the precise value of $\alpha(A)$ or $\chi(A)$ for a given set A in a Banach space. However, there are some criteria which hold in special settings. We describe two here; details may be found in Banas and Goebel, 1980.

Example 8.1 Consider the space $\mathscr{C}[0, 1]$. The modulus of continuity of a function $f \in \mathscr{C}[0, 1]$ is the function

$$w(f, \epsilon) = \sup\{|f(t) - f(s)|: t, s \in [0, 1], |t - s| < \epsilon\}.$$

For any bounded set $A \subset [0, 1]$, denote

$$w(A, \epsilon) = \sup\{w(f, \epsilon): f \in A\};$$

$$w_0(A) = \lim_{\epsilon \to 0^+} w(A, \epsilon).$$

It is possible to show that the function w_0 satisfies all the properties (a)–(i) and, moreover, that $\chi(A) = \frac{1}{2}w_0(A)$.

Example 8.2 Let X denote a Banach space which has a Schauder basis $\{e^i\}, i = 0, 1, 2, \ldots$. Then each element $x \in X$ has a unique representation

$$x = \sum_{i=0}^{\infty} \varphi_i(x)e^i$$

where φ_i are the basic functionals. Denote by S_n and R_n the operations

$$S_n x = \sum_{i=1}^{n} \varphi_i(x)e^i; \qquad R_n x = \sum_{i=n+1}^{\infty} \varphi_i(x)e^i, \qquad n = 0, 1, \ldots.$$

Since the family of linear operators $\{S_n, R_n\}$ is equibounded,

$$L = \lim_{n \to \infty} \sup |||R_n||| < +\infty,$$

where $|||\cdot|||$ denotes the operator norm (on the space of bounded linear operators on X). If A is a bounded subset of X, set

$$\mu(A) = \lim_{n \to \infty} \sup \|R_n A\|$$

where

$$\|R_n A\| = \sup\{\|R_n x\|: x \in A\}.$$

Then the function μ is another measure of noncompactness (satisfying (a)–(i)). It compares with the measure χ as follows:

$$L^{-1}\mu(A) \leqslant \chi(A) \leqslant \inf\{\|R_n A\|: n = 0, 1, \ldots\} \leqslant \mu(A). \tag{8.1}$$

In the special case $L = 1$ the measure μ coincides with the Hausdorff measure χ. This occurs in the space c_0, in all l^p spaces, $1 \leqslant p < \infty$, with the standard Schauder basis, and also in the spaces D_1, D_∞ of Day mentioned in Example 5.5. In general, it is well known that a subset $A \in \mathscr{B}$ in a space

X with a basis is precompact if and only if $\lim_{n\to\infty}\|R_nA\|=0$. The other properties of the measure of noncompactness are easy to check.

To prove (8.1), observe that $\chi(S_nA)=0$ for all n since S_nA is finite dimensional, hence compact. Since $R_nA \subset A+(-S_nA)$ and $A \subset R_nA+S_nA$, by virtue of (g) we have $\chi(A)=\chi(R_nA), n=0,1,\dots$. Now the obvious inequality $\chi(R_nA)\leqslant\|R_nA\|$ implies

$$\chi(A)\leqslant\inf\{\|R_nA\|:n=0,1,\dots\}.$$

Now assume $\chi(A)=\chi(R_nA)=r$. Then given $\epsilon>0$ there exists a finite set A_0 such that

$$A\subset B(A_0;r+\epsilon)=\bigcup_{a\in A_0}B(a;r+\epsilon).$$

Let $x\in A$ and select $a\in A_0$ such that $\|x-a\|\leqslant r+\epsilon$. Then

$$\|R_nx\|-\|R_na\|\leqslant\|R_n(x-a)\|\leqslant\|\|R_n\|\|\|x-a\|\leqslant\|\|R_n\|\|(r+\epsilon),$$

and since $\lim_{n\to\infty}\|R_nA_0\|=0$,

$$\limsup_{n\to\infty}\|R_nA\|\leqslant(r+\epsilon)\limsup_{n\to\infty}\|\|R_n\|\|.$$

This proves the first inequality of (8.1). The last inequality is trivial.

As with the modulus of convexity, it is possible to use the 'modulus of noncompact convexity' to scale Banach spaces.

Definition 8.3 The *modulus of noncompact convexity* of a Banach space X is the function $\Delta_X:[0,2]\to[0,1]$ defined by

$$\Delta_X(\epsilon)=\inf\{1-\mathrm{dist}(0,A)\}=\inf\left\{1-\inf_{x\in A}\{\|x\|\}\right\},$$

where the first infimum is taken over all convex subsets A of the unit ball with $\alpha(A)\geqslant\epsilon$.

The following implication is an immediate consequence of the definition:

$$A\text{ convex, }A\subset B(x;r),\text{ and }\alpha(A)\geqslant\epsilon r\Rightarrow\mathrm{dist}(x,A)\leqslant(1-\Delta_A(\epsilon))r.\quad(8.2)$$

The above of course is a counterpart of the implication of (5.2) which involves the modulus of convexity δ_X. It is formulated in terms of Kuratowski's measure, rather than some other measure of noncompactness, for two reasons. First, for infinite dimensional spaces, $\alpha(B(0;1))=2$, and this insures that Δ_X is well defined on $[0,2]$, a fact which facilitates a comparison of Δ_X and δ_X. Second, any set A with $\alpha(A) > \epsilon$ also has diam $(A) > \epsilon$, implying $\Delta_X(\epsilon) \geqslant \delta_X(\epsilon)$ for any $\epsilon \in [0, 2]$; hence $\epsilon_1(X) \leqslant \epsilon_0(X)$ if ϵ_1 is defined as follows:

Definition 8.4 The *coefficient of noncompact convexity* of X is the number

$$\epsilon_1(X) = \sup\{\epsilon \geqslant 0 : \Delta_X(\epsilon) = 0\}.$$

A space for which $\epsilon_1(X) = 0$ is said to be Δ-*uniformly convex*.

Obviously the class of Δ-uniformly convex spaces includes all uniformly convex spaces.

Unfortunately, as with the modulus δ_X, it is often difficult to give a precise formula for Δ_X, although it is possible to do so in special circumstances.

Example 8.3 Let $X = l^p(\mathbb{N})$, $1 < p < \infty$, and let $\{e^i\}$ be the standard basis for l^p. Notice that all the operators S_n and R_n discussed in Example 8.2 have norm 1; hence if A satisfies $\alpha(A) \geqslant \epsilon$,

$$\chi(A) = \lim_{n \to \infty} \|R_n A\| \geqslant \epsilon/2.$$

Now let A be any closed and convex subset of the unit ball in l^p for which $\alpha(A) \geqslant \epsilon$. Then A must contain a sequence $\{x_n\}$ for which $\lim_{n \to \infty} \inf \|R_n x_n\| \geqslant \epsilon/2$. Since l^p is reflexive we may assume $\{x_n\}$ converges weakly, say to $z \in A$. Let $\mu > 0$. Then for k sufficiently large, $\|R_k z\| < \mu$, and for $n \geqslant k$,

$$1 \geqslant \|x_n\|^p = \|S_k x_n\|^p + \|R_n x_n\|^p$$

$$\geqslant \|S_k x_n\|^p + (\epsilon/2)^p$$

and letting $n \to \infty$,

$$1 \geqslant \|S_k z\|^p + (\epsilon/2)^p$$

$$\geqslant \|z\|^p - \mu^p + (\epsilon/2)^p.$$

Since μ is arbitrary,

$$\|z\| \leqslant (1 - (\epsilon/2)^p)^{1/p};$$

consequently

$$\Delta_{l^p}(\epsilon) \geqslant 1 - (1 - (\epsilon/2)^p)^{1/p}.$$

Considering the set

$$\left\{ x = \sum_{i=1}^{\infty} \xi_i \, e^i : \|x\| \leqslant 1, \xi_1 \geqslant (1 - (\epsilon/2)^p)^{1/p} \right\}$$

we see that, in fact,

$$\varDelta_{l^p}(\epsilon) = 1 - (1 - (\epsilon/2)^p)^{1/p}.$$

For $p \geqslant 2$ this coincides with the known value of the modulus of convexity:

$$\varDelta_{l^p}(\epsilon) \equiv \delta_{l^p}(\epsilon).$$

However, for $1 < p < 2$,

$$\varDelta_{l^p}(\epsilon) > \varDelta_{l^2}(\epsilon) = \delta_H(\epsilon) > \delta_{l^p}(\epsilon).$$

In view of this, the functions $\varDelta_H$ and δ_X may differ and thus provide different scalings of Banach spaces. Moreover, it seems that Hilbert space is not 'the most uniformly rotund' with respect to the $\varDelta_X$ scaling.

Finally, we observe that for $p = 1$, $\varDelta_{l^1}(\epsilon) \equiv 0$. Indeed, the set

$$C = \text{conv}\{e^i\} = \left\{ x = \sum_{i=1}^{\infty} \xi_i \, e^i : \xi_i \geqslant 0, \sum_{i=1}^{\infty} \xi_i = 1 \right\}$$

is contained in the unit ball, with $\alpha(C) = 2$. On the other hand,

$$\lim_{p \to 1} \varDelta_{l^p}(\epsilon) = \epsilon/2 \neq \varDelta_{l^1}(\epsilon).$$

We shall return to this observation later.

Although the l^p spaces are all uniformly convex, our real motivation for introducing the modulus $\varDelta_X$ is to facilitate the study of a wider class of spaces. The advantage of the modulus $\varDelta_X$ lies in the fact that its values are not affected by 'flat compact sets which lie near the unit sphere'. The following example illustrates this.

Example 8.4 We consider the two Day spaces D_1 and D_∞ introduced in Example 5.5:

$$D_1 = \left\{ x = \{x^n\}_{n=1}^{\infty} : x^n \in l_n^1, \left(\sum_{i=1}^{\infty} (|x^i|_i^1)^2 \right)^{1/2} = \|x\|_{D_1} < \infty \right\};$$

$$D_\infty = \left\{ x = \{x^n\}_{n=1}^{\infty} : x^n \subset L_n^\infty, \left(\sum_{i=1}^{\infty} (|x^i|_i^\infty)^2 \right)^{1/2} = \|x\|_{D_\infty} < \infty \right\}.$$

Each of these spaces has a standard basis $\{e^{n,k}\}$ where $e^{n,k}$ is the kth basis unit vector in the factor spaces l_1^n, L_n^∞ of the corresponding spaces D_1, D_∞. Considering the sequences $\{S_{n,n}\}$ and $\{R_{n,n}\}$ of projection and remainder operators onto the first n terms of the respective direct sums of $l_k^1, L_k^\infty, k =$

$1, \ldots, n$, and repeating the calculations of Example 8.3, we obtain

$$\varDelta_{D_1}(\epsilon) = \varDelta_{D_\infty}(\epsilon) = \delta_H(\epsilon) = 1 - (1 - (\epsilon/2)^2)^{1/2}.$$

However, the spaces D_1, D_∞ are not uniformly convex, nor even superreflexive. Thus $\delta_{D_1}(\epsilon) = \delta_{D_\infty}(\epsilon) = 0$ for every $\epsilon \in (0, 2)$, even under equivalent renorming. Moreover, since D_1 and D_∞ contain, respectively, isometric copies of l_n^1 and L_n^∞ for every n, and since the normal structure coefficients of the later spaces satisfy

$$N(l_n^1) \geqslant 1 - (1/n), \qquad N(L_n^\infty) \geqslant 1 - (1/n),$$

the spaces D_1 and D_∞ do not have uniform normal structure. (The above inequalities may easily be checked by taking the convex envelope of the unit vectors in l_n^1 and L_n^∞.)

The above example shows that $\varDelta$-uniform convexity does not imply uniform normal structure. However, as we shall see later, it does imply normal structure.

Before turning to the main results of this chapter we discuss one more example.

Example 8.5 The spaces D_1 and D_∞ fit into a more general context. In 1969, J. P. Gossez and E. Lami Dozo (1969) introduced a condition on a space with a basis which implies both normal structure for reflexive spaces and weak* normal structure for dual spaces. A Banach space X with basis $\{e^i\}$ is said to satisfy the Gossez–Lami Dozo (G–L) *condition* if there exists an increasing sequence $\{k_n\}$ of integers such that for every $\epsilon > 0$ there exists $\delta > 0$ such that for every $x \in X$ and $n \in \mathbb{N}$:

$$\left. \begin{array}{l} \|S_{k_n} x\| = 1 \\ \|R_{k_n} x\| \geqslant \epsilon \end{array} \right\} \Rightarrow \|x\| > 1 + \delta.$$

It is easy to see that all the l^p spaces, $1 \leqslant p < \infty$, as well as D_1 and D_∞ satisfy the G–L condition and, since l^1 is not reflexive, this condition does not imply reflexivity.

We claim that all reflexive spaces which satisfy the G–L condition are $\varDelta$-uniformly convex. To see this, let X be such a space and set

$$r_n(c) = \inf \|x\| - 1,$$

where $c \geqslant 0$ and the infimum is taken over all $x \in X$ which satisfy $\|S_{k_n} x\| = 1$ and $\|R_{k_n} x\| \geqslant c$. Also set

$$r(c) = \inf\{r_n(c) : n = 1, 2, \ldots\}.$$

Clearly $r(0)=0$ and $r(c)>0$ for $c>0$, and it is known that r is continuous and nondecreasing.

Now let A be a closed, convex subset of the unit ball of X with $\alpha(A)\geqslant\epsilon$. Then $\chi(A)\geqslant\epsilon/2$, and in view of (8.1) there exists a sequence of points $\{x_n\}\subset A$ such that

$$\liminf_{n\to\infty}\|R_{k_n}x_n\|\geqslant\epsilon/2.$$

Furthermore, since X is reflexive, we may assume w-$\lim_{n\to\infty}x_n=z\in A$. It is not difficult to check that

$$z=\text{w-}\lim_{n\to\infty}S_{k_n}x_n\qquad\text{and}\qquad\text{w-}\lim_{n\to\infty}R_{k_n}x_n=0.$$

Consider the sequence $\{y_n\}=\{x_n/\|S_{k_n}x_n\|\}$. Then $\|S_{k_n}y_n\|=1$, and $\|R_{k_n}y_n\|\geqslant\epsilon/(2\|S_{k_n}x_n\|)$ implying

$$\|y_n\|=\frac{\|x_n\|}{\|S_{k_n}x_n\|}\geqslant1+r(\epsilon/(2\|S_{k_n}x_n\|)).$$

Since $\|x_n\|\leqslant1$,

$$\|S_{k_n}x_n\|(1+r(\epsilon/(2\|S_{k_n}x_n\|)))\leqslant1$$

and finally, if we put $a=\liminf\|S_{k_n}x_n\|$, $\|z\|\leqslant a$ and

$$\|z\|(1+r(\epsilon/2a))\leqslant1. \tag{8.3}$$

This shows that z is bounded away from the unit sphere, yielding $\Delta_X(\epsilon)>0$ for $\epsilon>0$. Note also that (8.3) provides an implicit estimate for $\Delta_X(\epsilon)$.

One of many interesting results of R. C. James is the following characterization of nonreflexive spaces (see James, 1964a):

Theorem 8.1 *A Banach space X is nonreflexive if and only if for any $t\in(0,1)$ there exists sequences $\{x_n\}$ in the unit ball of X and $\{f_n\}$ in the unit ball of X^* such that for $i,j=1,2,\ldots$,*

$$f_j(x_i)=\begin{cases}t & \text{if } j\leqslant i\\0 & \text{if } j>i.\end{cases}$$

Note that under the assumptions of the theorem, if $A=\text{conv}\{x_1,x_2,\ldots\}$ then for any $z\in A$, $\|z\|\geqslant t$, and in view of $\|x_j-x_i\|\geqslant f_j(x_j-x_i)=t$ for $j>i$, $\alpha(A)\geqslant t$. Hence

$$\Delta_X(t)\leqslant1-t$$

and by monoticity of $\Delta_{X,\,\epsilon_1}(X) \geq 1$. This proves the following:

Theorem 8.2 *If $\epsilon_1(X) < 1$, then X is reflexive.*

Consequently, all Δ-uniformly convex spaces are reflexive. However, we note that the condition $\epsilon_1(X) < 1$ is not necessary for reflexivity:

Example 8.6 The space X_λ defined as l^2 equivalently renormed by taking for $x = \{x_1, x_2, \dots\} \in l^2$ and fixed $\lambda \geq 1$,

$$\|x\|_\lambda = \max\{\lambda|x_1|, \|x\|_{l^2}\},$$

has $\epsilon_1 = 2(1 - \lambda^{-2})^{1/2}$. (Thus $\epsilon_1 \to 2$ as $\lambda \to \infty$.)

A more interesting fact in the context of the present study is the following:

Theorem 8.3 *If $\epsilon_1(X) < 1$, then X has normal structure.*

Proof Suppose X is a Banach space for which $\epsilon_1(X) < 1$ and which does not have normal structure, and let K be a bounded, closed, convex, diametral subset of X which consists of more than one point. By normalizing, we may assume diam $K = 1$. For any functional $f \in X^*$ with $\|f\| = 1$ and any $d \in [0, 1)$, consider the set (an open half-space):

$$U(f, d) = \{x \in X : f(x) > d\}.$$

Observe that

$$d \leq \inf\{\|x\| : x \in U(f, d) \cap B(0; 1)\} \leq 1 - \Delta_X(\alpha(U(f, d) \cap B(0; 1))).$$

Since $\epsilon_1(X) < 1$ there exists $\xi > 0$ such that for d sufficiently near 1,

$$\alpha(U(f, d) \cap B(0; 1)) < 1 - \xi$$

for all norm one functionals f. Now consider the family of all sets of the form

$$V(x, f, d) = x + U(f, d), \qquad x \in K.$$

Such sets are weakly open and, since K is diametral, K is contained in the union of all $V(x, f, d)$. Indeed, if $x \in K$ there exists $y \in K$ such that $\|x - y\| > d$, so $x \in V(y, f, d)$ if f is any norm one functional for which $f(x - y) = \|x - y\|$.

Now, since X is reflexive and K weakly compact, K is covered by a finite number of the (weakly open) sets $V(x, f, d)$, say

$$K \subset \bigcup_{i=1}^{n} V(x_i, f_i, d).$$

Hence

$$\alpha(K) \leqslant \max_{i} \alpha(V(x_i, f_i, d) \cap K)$$

$$\leqslant \max_{i} \alpha(V(x_i, f_i, d) \cap B(x_i; 1))$$

$$\leqslant 1 - \xi.$$

This contradicts the easily shown fact that for diametral sets $K, \alpha(K) = $ diam K and, in this instance, $\alpha(K) = 1$.

Although we do not conclude in the above theorem that X has uniform normal structure, the type of normal structure obtained is stable in the following sense:

Lemma 8.1 *Let* $\|\cdot\|_1$ *and* $\|\cdot\|_2$ *be two equivalent norms on a Banach space* X; *say* $\alpha\|x\|_1 \leqslant \|x\|_2 \leqslant \beta\|x\|_2, x \in X$, *and set* $k = \beta/\alpha$. *Let* X_1 *and* X_2 *denote, respectively,* $(X, \|\cdot\|_1)$ *and* $(X, \|\cdot\|_2)$. *Then*

$$\Delta_{X_2}(\epsilon) \geqslant 1 - k(1 - \Delta_{X_1}(\epsilon/k)). \tag{8.4}$$

Consequently, if $\epsilon_1(X_1) < 1$ *and* $k(1 - \Delta_{X_1}(1/k)) < 1$, *then* $\epsilon_1(X_2) < 1$.

The proof of the above is routine and follows that of Theorem 6.3.

Therefore normal structure for spaces X satisfying $\epsilon_1(X) < 1$ is in a sense stable, although in a weaker sense than for spaces satisfying $\epsilon_0(X) < 1$. Specifically, we have no analogue of Theorem 6.4; as far as we know, the question of whether any reflexive space X has an equivalent norm satisfying $\epsilon_1(X) < 1$ remains open. Nor is it known if such norms are dense in the family of all renormings of X. Finally, it is also unknown if spaces for which $\epsilon_1(X) < 1$ have an equivalent Δ-uniformly convex norm.

A closer analysis of the proof of Theorem 8.3 leads to an estimate analogous to those following Theorems 6.1 and 6.1'.

Let K be weakly compact and let $C(K)$ and $r = r(K)$ denote, respectively, the Chebyshev center and Chebyshev radius of K. Take $0 < \epsilon < r$, and for $x \in K$ and $f \in X^*$ with $\|f\| = 1$, consider the family of half-spaces $\{V(x, f, r - \epsilon)\}$. Since for any $y \in C(K)$ there exists $x \in K$ such that $\|x - y\| > r - \epsilon$, the same reasoning as given in the proof of Theorem 8.3 shows that a finite collection of such sets covers $C(K)$, and among them there is one, say $V(\bar{x},\bar{f},r-\epsilon)$, which satisfies

$$\alpha(V(\bar{x},\bar{f},r-\epsilon)\cap C(K))=\alpha(C(K)).$$

Since $C(K)\subset B(x;r)$,

$$r-\epsilon\leqslant\left\{1-\varDelta\left[\frac{\alpha(C(K))}{r}\right]\right\}r$$

and letting $\epsilon\rightarrow0$ we conclude

$$\alpha(C(K))\leqslant\epsilon_1(X)r(K). \tag{8.5}$$

When $\epsilon_1(X)<1$ the above provides an alternate proof of Theorem 8.3. In particular, it shows that *convex* sets in $\varDelta$-uniformly convex spaces have compact Chebyshev centers (cf., van Dulst and Sims, 1983).

It is possible to extend the methods presented above to certain nonreflexive dual spaces. Suppose X is a dual space with a dual norm, i.e., suppose $X=Y^*$ with

$$\|x\|=\sup\{x(y):y\in Y,\|y\|\leqslant1\}, \qquad x\in X.$$

The unit ball in X is weak* compact (Banach–Alaoglu Theorem) and it is possible to define a new *modulus of noncompact convexity*, $\varDelta_X^*$, by setting

$$\varDelta_X^*(\epsilon)=\inf\{1-\operatorname{dist}(0,A)\}, \qquad \epsilon\in[0,2],$$

where the infimum is taken over all weak* compact, convex subsets A of $B(0;1)$ with $\alpha(A)\geqslant\epsilon$. Also set $\epsilon_1^*(X)=\sup\{\epsilon:\varDelta_X^*(\epsilon)=0\}$.

Obviously $\varDelta_X^*(\epsilon)\geqslant\varDelta_X(\epsilon)$ for any dual space X, and in some instances a strict inequality holds.

Example 8.7 Consider $X=l^1$ as a dual space of c_0. In this instance weak* convergence in the unit ball B is just coordinate-wise convergence. By repeating the arguments used in determining $\varDelta_{l^p}(\epsilon)$ (Example 8.3) we obtain

$$0=\varDelta_{l^1}(\epsilon)<\varDelta_{l^1}^*(\epsilon)=\epsilon/2=\lim_{p\to1}\varDelta_{l^p}(\epsilon).$$

A repetition of the proof of the previous theorem yields:

Theorem 8.4 *If X is a dual Banach space with dual norm and if $\epsilon_1^*(X)<1$, then X has weak* normal structure (weak* compact, convex sets of positive diameter are nondiametral).*

In the case of dual spaces the question of stability of normal structure is not clear. Not all equivalent norms on a dual space are dual norms; hence weak* compactness may not be preserved under renorming. However, within the context of equivalent dual norms, one has stability in the sense of (8.4).

We note also that the result of Example 8.5 cannot be translated directly into a dual space context. Suppose $X = Y^*$ is a dual space having a basis $\{e^n\}$ satisfying the (G–L) condition with respect to its natural dual norm. By repeating, step by step, the reasoning of Example 8.5 one obtains the implication (G–L) implies $\epsilon_1^*(X) = 0$ only if weak* convergence coincides with boundedness and coordinate-wise convergence. This is, of course, the case when l^1 is treated as the dual of c_0 (recall Example 4.1). However, in general this is not the case.

Example 8.8 Consider the space c of all convergent sequences of real numbers with supremum norm. The simplest basis for c consists of the constant vector $e = (1, 1, \ldots)$ along with the unit vectors $e^n = \{\delta_{in}\}$. For any $x = \{\xi_i\} \in c$ the expansion with respect to this basis is

$$x = \lim x \, e + \sum_{n=1}^{\infty} (\xi_i - \lim x) \, e^n$$

(where $\lim x = \lim_{i \to \infty} \xi_i$). Any linear functional $y \in c^*$ satisfies

$$y(x) = \langle x, y \rangle = \lim x \cdot y(e) + \sum_{n=1}^{\infty} (\xi_i - \lim x) \, y(e^n)$$

$$= \lim x \left(y(e) - \sum_{n=1}^{\infty} y(e^n) \right) + \sum_{n=1}^{\infty} \xi_i y(e^n).$$

Since y also acts on c_0 as a closed subspace of c, the sequence $\{y(e^n)\}$ must be in l^1. Thus the biorthogonal system of functionals for our basis in c consists of the nontrivial functional 'lim' along with all coordinate functionals ξ_i.

We conclude that any $y \in c^*$ can be represented in $l^1(N_0)$ as a sequence $y = \{\eta_i\}, i = 0, 1, 2, \ldots$, such that

$$y(x) = \langle x, y \rangle = \eta_0 \lim x + \sum_{n=1}^{\infty} \xi_i \eta_i.$$

Obviously, $|y(x)| \leqslant \sup |\xi_i| \left(\sum_{n=0}^{\infty} |\eta_n| \right)$ and this implies $\| y \|_{c^*} \leqslant \| y \|_{l^1}$. On the other hand, for

$$x^k = \begin{cases} \operatorname{sgn} \eta_n & n = 0, 1, 2, \ldots, k \\ 1 & n = k+1, k+2, \ldots, \end{cases}$$

we have

$$y(x^k) = \sum_{n=0}^{k} |\eta_n| + \sum_{n=k+1}^{\infty} \eta_n;$$

hence $y(x^k) \to \|y\|_{l^1}$ as $k \to \infty$. Thus $\|y\|_{c*} = \|y\|_{l^1}$.

The above shows that $c*$ is isometric with l^1. However, the weak* topology induced in l^1 by c is different than that induced by c_0. For example, w*-$\lim_{n \to \infty} e^n = \lim$ (which is represented in $l^1 = c*$ as e_0) is a nontrivial functional. Also, the subspace E of l^1 defined by

$$E = \{y \in l^1 : y(e) = \langle e, y \rangle = 0\}$$

$$= \left\{ y = \{\eta_i\} \in l^1 : \sum_{i=1}^{\infty} \eta_i = 0 \right\}$$

is weak* closed in $l^1 = c*$. Consequently, the affine variety $E_1 = \{y \in l^1 : y(e) = 1\}$ is also weak* closed, so the set $E_1 \cap B$ is weak* compact in $l^1 = c*$. However,

$$E_1 \cap B = \text{conv}\{e^n : n = 0, 1, \ldots\}$$

is diametral and lacks the fixed point property (f.p.p.).

In summary, l^1 with the weak* topology induced by its predual c satisfies the (G–L) condition, yet it fails to have weak* normal structure and the weak* fixed point property.

On the other hand, l^1 has the weak* f.p.p. relative to the weak* topology induced by c_0, and the stability of the f.p.p. may be considered in this context. This problem has been completely solved. Recently, using nonstandard techniques which we discuss in Chapter 14, Khamsi (1987) has shown that l^1 with an equivalent norm for which $k = \beta/\alpha < 2$ has weak* normal structure. This establishes Soardi's assertion (1982) that weak* compact, convex sets in such spaces have the f.p.p.

The following example (due to Lim (1980)) shows that the assumption $k < 2$ cannot be improved.

Example 8.9 Let c_0 be the space of all sequences $x = \{x_1, x_2, \ldots\}$ converging to 0 equipped with the standard norm: $\|x\|_\infty = \sup\{|x_i| : i = 1, 2, \ldots\}$, and let l^1 have its standard norm: $\|x\|_{l^1} = \sum_{i=1}^{\infty} |x_i|$. For any $x = \{x_1, x_2, \ldots\} \in c_0$ let

$$x^+ = \{x_i^+\} \qquad \text{and} \quad x^- = \{x_i^-\},$$

where $x_i^+ = \max\{x_i, 0\}$ and $x_i^- = \min\{x_i, 0\}$, and renorm c_0 by setting

$$|x|_0 = \|x^+\|_\infty + \|x^-\|_\infty.$$

Then

$$\|x\|_\infty \leqslant |x|_0 \leqslant 2\|x\|_\infty.$$

The dual of $(c_0, |\cdot|_0)$ is the space $(l^1, |\cdot|_1)$ where

$$|f|_1 = \max\{\|f^+\|_{l^1}, \|f^-\|_{l^1}\}, \qquad f = \{f_1, f_2, \ldots\} \in l^1,$$

and $|\cdot|_1$ is the dual norm to $|\cdot|_0$.

Now consider the set

$$K = \left\{ \{x_i\} \in l^1 : x_i \geqslant 0, \ \sum_{i=1}^\infty x_i \leqslant 1 \right\}.$$

Then K is a weak* compact convex set in l^1, and the mapping $T: K \to K$ defined by

$$Tx = \left(1 - \sum_{i=1}^\infty x_i, x_1, x_2, \ldots \right)$$

has Lipschitz constant 2 with respect to the original l^1 norm. However T is an isometry with respect to $|\cdot|_1$, and obviously T is fixed point free. Notice also that for any $x \in K$,

$$\tfrac{1}{2}\|x\|_{l^1} \leqslant |x|_1 \leqslant \|x\|_{l^1}.$$

9

Sequential approximation techniques for nonexpansive mappings

As previous examples illustrate, neither iterative sequences nor approximate fixed point sequences for nonexpansive mappings typically converge, at least in the strong sense. However, in certain instances, these sequences may converge in the weak (or weak*) topology to fixed points, or in some other way determine invariant sets which contain fixed points.

Throughout this chapter we shall, as usual, assume that K is a nonempty, closed and convex subset of a Banach space X but, in general, we shall not assume K is bounded.

Suppose $T: K \to K$ is nonexpansive and fix $x_0 \in K$. We begin by considering the iterative sequence $\{x_n\} = \{T^n x_0\}$. The set of points $O(x_0) = \{x_n : n = 0, 1, \ldots\}$ is called the *orbit* of x_0 under T, and its closure is called the *closed orbit*. Since T is nonexpansive, if $O(x_0)$ is bounded for at least one $x_0 \in K$ then all other orbits $O(x)$, $x \in K$, are bounded. Thus the nonexpansive self-mappings of K fall into two categories: those with bounded orbits and those with unbounded orbits. Obviously all nonexpansive mappings having fixed points have bounded orbits. (Less obvious is the fact that in a finite dimensional Banach space $O(x_0)$ is bounded for a nonexpansive mapping T whenever $\{x_n\}$ has a convergent subsequence (see Roehrig and Sine, 1981).)

In our previous examples of fixed point free, nonexpansive mappings none of the iterative sequences contain convergent subsequences. However Edelstein has shown (1964) that even if such convergent subsequences exist, a nonexpansive mapping may remain fixed point free.

Example 9.1 Let l^2 be the space of square summable *complex* sequences with the classical norm:

$$\| x \| = \| \{x_1, x_2, \ldots\} \| = \left(\sum_{n=1}^{\infty} |x_n|^2 \right)^{1/2},$$

and define $T: l^2 \to l^2$ by taking $Tx = y$ where $y = \{y_n\}$ with

$$y_n = [\exp(2\pi i/n!)](x_n - 1) + 1, \qquad n = 1, 2, \ldots.$$

Then

$$\sum_{n=1}^{\infty} |y_n|^2 \leqslant \sum_{n=1}^{\infty} |x_n|^2 + 4 \sum_{n=1}^{\infty} |x_n| \sin \frac{\pi}{n!} + 4 \sum_{n=1}^{\infty} \sin^2 \frac{\pi}{n!}$$

$$\leqslant \|x\|^2 + 4\|x\| \left(\sum_{n=1}^{\infty} \sin^2 \frac{\pi}{n!} \right)^{1/2} + 4 \sum_{n=1}^{\infty} \sin^2 \frac{\pi}{n!}$$

$$< \infty,$$

and if $y^1 = Tx^1$ and $y^2 = Tx^2$, then

$$\sum_{n=1}^{\infty} |y_n^1 - y_n^2|^2 = \sum_{n=1}^{\infty} |x_n^1 - x_n^2|^2$$

Thus T is an isometry on l^2. Clearly the only sequence which is invariant under T is $\{1, 1, \ldots\}$, and since this point is not in l^2, T is fixed point free. On the other hand, for $(0) = \{0, 0, \ldots\}$,

$$T(0) = \{1 - \exp(2\pi i/n!)\},$$

$$T^2(0) = \{1 - \exp(2(2\pi i/n!))\}, \ldots, T^k(0) = \{1 - \exp(k(2\pi i/n!))\}$$

and

$$\|T^k(0)\|^2 = \sum_{n=1}^{\infty} |1 - \exp(k(2\pi i/n!))|^2 = 4 \sum_{n=1}^{\infty} \sin^2(\pi k/n!).$$

Thus if $p_k = k!$ and $q_k = \frac{1}{2} \sum_{j=1}^{k} (2^j k)!$, routine calculations show that

$$0 = \lim_{k \to \infty} \|T^{p_k}(0)\| < \lim_{k \to \infty} \|T^{q_k}(0)\| = +\infty.$$

Example 9.2 Consider the space c of all convergent sequences $x = \{x_n\}$ of complex numbers with $|x| = \sup |x_n|$. Define $T: c \to c$ by setting $Tx = y$ where $y = \{y_n\}$ is defined by

$$y_n = [\exp(2\pi i/n!)] x_n - (-1)^n (\exp(2\pi i/n!) - 1), \qquad n = 1, 2, \ldots.$$

Then T is an isometry on c, and for $(0) = \{0, 0, \ldots\}$:

$$T(0) = \{(-1)^n [1 - \exp(2\pi i/n!)]\}, \qquad T^2(0) = \{(-1)^n [1 - \exp(2(2\pi i/n!))]\},$$

and, generally,

$$T^k(0) = \{(-1)^n [1 - exp(k(2\pi i/n!))]\}.$$

Thus $\{T^k(0)\}$ remains bounded in c, and it is easy to see that $\lim_{k \to \infty} T^{k!}(0) = 0$. On the other hand, T is fixed point free since the only sequence which is a candidate for a fixed point is $\{x_n\}=\{(-1)^n\}$ which is not an element of c. Obviously the sequence $\{T^k(0)\}$ does not converge.

The following indirect method for studying fixed point properties (f.p.p.) of nonexpansive mappings via iterative and approximate sequences was introduced by M. Edelstein in 1972. For $x \in X$ and bounded $\{x_n\} \subset X$, define the *asymptotic radius of* $\{x_n\}$ at x as the number

$$r(x, \{x_n\}) = \lim_{n \to \infty} \sup \|x - x_n\|. \tag{9.1}$$

If $\{x_n\}$ is fixed, (9.1) defines a function on X having nonnegative real values. This function is easily seen to have the following properties:

(a) For all $x \in X$, $r(x, \{x_n\})=0 \Leftrightarrow \lim_{n \to \infty} x_n = x$.

(b) For all $x, y \in X$, $|r(x, \{x_n\}) - r(y, \{x_n\})| \leqslant \|x - y\|$.

(c) For all $x, y \in X$ and $\alpha, \beta \geqslant 0$ with $\alpha + \beta = 1$,

$r(\alpha x + \beta y, \{x_n\}) \leqslant \alpha r(x, \{x_n\}) + \beta r(y, \{x_n\})$.

Therefore $r(x, \{x_n\})$ is a nonnegative, continuous, and convex function of x; hence:

(d) For any $a \geqslant 0$, the 'level set'

$$A_a(\{x_n\}) = \{x : r(x, \{x_n\}) \leqslant a\}$$

is closed and convex. Of course if $\{x_n\}$ does not converge, $A_a(\{x_n\})$ will be empty for sufficiently small a.

Now let K be a given nonempty, closed subset of X. The *asymptotic radius of* $\{x_n\}$ *in* K is the number

$$r(K, \{x_n\}) = \inf\{r(x, \{x_n\}) : x \in K\}. \tag{9.2}$$

The corresponding level sets are defined for $\mu \geqslant 0$ by

$$A_\mu(K, \{x_n\}) = \{x \in K : r(x, \{x_n\}) \leqslant r(K, \{x_n\}) + \mu\}. \tag{9.3}$$

The set $A_0(K, \{x_n\})$ is called *the asymptotic center of* $\{x_n\}$ *in* K. Usually we write $A(K, \{x_n\})$, or even $A(\{x_n\})$ when K is fixed, for $A_0(K, \{x_n\})$.

Obviously $A_\mu(K, \{x_n\}) \neq \emptyset$ for all $\mu > 0$ and $A_{\mu'}(K, \{x_n\}) \subset A_\mu(K, \{x_n\})$ when $\mu' < \mu$, but it may be the case that

$$A(K, \{x_n\}) = A_0(K, \{x_n\}) = \bigcap_{\mu > 0} A_\mu(K, \{x_n\}) = \emptyset. \tag{9.4}$$

We note that the asymptotic center of $\{x_n\}$ in K is also given by

$$A(K, \{x_n\}) = \bigcap_{\mu > 0} \left\{ \bigcup_{n=1}^{\infty} \bigcap_{i=n}^{\infty} B(x_i; r(K, \{x_n\}) + \mu) \right\} \cap K. \qquad (9.5)$$

The following obvious facts are fundamental to our purpose.

Lemma 9.1 *Let $\{x_n\}$ be a sequence in X and K a nonempty subset of X.*

(a) *If K is weakly compact, then $A(K, \{x_n\}) \neq \emptyset$.*
(b) *If K is convex, then $A(K, \{x_n\})$ is convex.*

Note that (9.5) characterizes $A(K, \{x_n\})$ as a descending family of weakly closed sets; thus (a) is immediate. (Also (b) is a consequence of the fact that any continuous convex function (in this instance $r(\cdot, \{x_n\})$) is lower semicontinuous with respect to the weak topology and thus achieves its infimum.

The significance of the asymptotic center concept in the theory of non-expansive mappings evolves from the following two lemmas.

Lemma 9.2 *Let X be a Banach space with $K \subset X$, and suppose $T: K \to K$ is nonexpansive with bounded orbits. For $x_0 \in K$, set $\{x_n\} = \{T^n x\}$. Then for each $\mu > 0$, $A_\mu(K, \{x_n\})$ is invariant under T.*

Proof It suffices to observe that for any $z \in K$,

$$r(Tz, \{x_n\}) = \limsup_{n \to \infty} \| Tz - T^n x_0 \|$$
$$\leqslant \limsup_{n \to \infty} \| z - T^{n-1} x_0 \|$$
$$= r(z, \{x_n\}).$$

Lemma 9.3 *Let T and K be as in Lemma 9.2 and suppose $\{y_n\}$ is an approximate fixed point sequence for T, i.e., suppose $\lim_{n \to \infty} \| y_n - Ty_n \| = 0$. Suppose further that $\{y_n\}$ is bounded. Then for each $\mu > 0$, $A_\mu(K, \{y_n\})$ is invariant under T.*

Proof The conclusion is immediate from the following:

$$r(Tz, y_n) = \limsup_{n \to \infty} \| Tz - y_n \|$$
$$\leqslant \limsup_{n \to \infty} (\| Tz - Ty_n \| + \| Ty_n - y_n \|)$$
$$\leqslant \limsup_{n \to \infty} \| z - y_n \|$$
$$= r(z, \{y_n\}).$$

The preceding observations may be summarized as follows:

Theorem 9.1 *Let K be a nonempty, closed and convex subset of a Banach space X and let $T: K \to K$ be nonexpansive. Then the following are equivalent:*

(a) *T has at least one bounded orbit.*
(b) *All orbits of T are bounded.*
(c) *K contains a nonempty, bounded, closed, convex, T-invariant subset.*
(d) *K contains a bounded approximate fixed point sequence.*

We now return to our conventional setting with K denoting a weakly compact convex subset of X and $T: K \to K$ nonexpansive.

Conclusion (c) of Theorem 9.1, which is a direct consequence of Lemmas 9.1 and 9.2, suggests the following procedure for attempting to locate fixed points of T. Select $x_0 \in K$, take $\{x_n\} = \{T^n x_0\}$, and set $K_1 = A(K, \{x_n\})$. Then K_1 is a nonempty, T-invariant, weakly compact, convex subset of K. By selecting a point in K_1 the procedure may be repeated to obtain a subset K_2 of K_1 which again is a nonempty, T-invariant, weakly compact and convex set. Continuing the procedure it is possible to obtain a decreasing sequence $K_1 \supset K_2 \supset K_3 \ldots$ of nonempty, weakly compact, convex, T-invariant sets. It will be shown that in many instances this procedure stabilizes and either yields the fixed point set or simplifies the problem. Not surprisingly, the behavior of the sequence $\{K_n\}$ depends deeply on the geometrical structure of X.

We should remark that the above procedure may be applied to an unbounded set K if K is locally weakly compact (i.e., the intersection of K with any ball is weakly compact) and if $T: K \to K$ has bounded orbits. Also, the same procedure may be followed using any approximate fixed point sequence instead of the iterative sequence.

There is a connection between the asymptotic center of a sequence and the modulus of convexity of the space.

Theorem 9.2 *Let K be a closed and convex subset of a Banach space X, and let $\{x_n\}$ be a bounded sequence in X. Then*

$$\text{diam } A(K, \{x_n\}) \leqslant \epsilon_0(X) r(K, \{x_n\}).$$

Proof If $A(K, \{x_n\})$ is either empty or a singleton there is nothing to prove, so we suppose $d = \text{diam } A(K, \{x_n\}) > 0$. Take $\mu \in (0, d)$ and select $u, v \in A(K, \{x_n\})$ so that $\|u - v\| \geqslant d - \mu$. Since

$$r(u, \{x_n\}) = \limsup_{n \to \infty} \|u - x_n\| = r(K, \{x_n\}),$$

$$r(v, \{x_n\}) = \limsup_{n \to \infty} \|v - x_n\| = r(K, \{x_n\}),$$

if $z = \frac{1}{2}(u+v)$, then $z \in A(K, \{x_n\})$ and a standard property of the modulus of convexity (see (5.3)) yields

$$r(K, \{x_n\}) = r(z, \{x_n\}) = \limsup_{n \to \infty} \| x_n - \tfrac{1}{2}(u+v) \|$$

$$\leqslant \left(1 - \delta \left(\frac{d-\mu}{r(K, \{x_n\})} \right) \right) r(K, \{x_n\}),$$

which in turn implies

$$d - \mu \leqslant \epsilon_0(X) r(K, \{x_n\}).$$

Since $\mu > 0$ is arbitrary, the proof is complete.

If $\epsilon_0(X) < 1$ then the procedure described after the statement of Theorem 9.1 leads directly to a fixed point. Specifically, if K is a nonempty, closed and convex subset of a Banach space X for which $\epsilon_0(X) < 1$, and if $T : K \to K$ is nonexpansive with bounded orbits, then for any $x_0 \in K$ the set $K_1 = A(K, T^n\{x_0\})$ is a nonempty, T-invariant, bounded, closed, convex set. Select $x_1 \in K_1$, set $K_2 = A(K_1, \{T^n x_1\})$, and continue the procedure to obtain a descending sequence $\{K_n\}$ of nonempty, T-invariant, *weakly compact* subsets of K. (Recall: $\epsilon_0(X) < 1 \Rightarrow X$ reflexive.) In view of Theorem 9.2,

$$\operatorname{diam} K_n \leqslant (\epsilon_0(X))^n r(K, \{T^n x_0\}). \tag{9.6}$$

Thus $K_\infty = \bigcap_{n=1}^\infty K_n$ consists of exactly one point which is fixed under T.

As a very special case of the above, if X is uniformly convex then $\epsilon_0(X) = 0$ and K_1 itself is a singleton which is fixed under T.

As before, approximate fixed point sequences instead of iterative sequences may be used in any stage of the above procedure. The estimate (9.6) remains valid.

A complete analogue of Theorem 9.2 (or formula 8.5) does not exist in terms of the modulus of noncompact convexity. However a partial analogue is known.

Let K be a closed, convex subset of a reflexive space X and let $\{x_n\}$ be a bounded sequence in K. Let $r = r(K, \{x_n\})$ and $A = A(K, \{x_n\})$, and select $0 < \epsilon < r$. Choose a subsequence $\{y_n\}$ of $\{x_n\}$ which converges weakly, say to $z \in K$, and which satisfies $\operatorname{sep}\{y_n\} \geqslant r - \epsilon$. Now let $x \in A$, and consider the sequence of sets $C_i = \overline{\operatorname{conv}}\{y_i, y_{i+1}, \ldots\}$, $i = 1, 2, \ldots$. Then for sufficiently large indices i, $C_i \subset B(x; r+\epsilon)$. Since $\alpha(C_i) \geqslant r - \epsilon$, there is a point $u_i \in C_i$ such that

$$\| x - u_i \| \leqslant \left[1 - \Delta \left(\frac{r-\epsilon}{r+\epsilon} \right) \right] (r+\epsilon).$$

In view of the fact that the above holds for all $x \in A$, we have the following. (Recall that r_K denotes the Chebyshev radius relative to K.)

Theorem 9.3 *Let K be a closed, convex subset of a reflexive Banach space and let $\{x_n\}$ be a bounded sequence in K. Then*

$$r_K(A(K, \{x_n\})) \leqslant (1 - \varDelta(1^-))r(K, \{x_n\}). \qquad (9.7)$$

Note that if $\epsilon_1(X) < 1$, the inequality (9.7) again yields a stabilizing sequence of T-invariants sets and in turn a fixed point of T. On the other hand, the inequality (9.7) is not as sharp as its analogue (8.5), and in particular it does not allow one to conclude that asymptotic centers of sequences in $\varDelta$-uniformly convex spaces are compact. In fact this in general is not the case, as the following example due to S. Prus (unpublished) shows:

Example 9.3 Let X be the space l^2 re-normed as follows. For $x = \Sigma_{i=1}^{\infty} x_i e_i$ ($\{e_i\}$ denotes the standard basis in l^2) set

$$\|x\| = \max_n \left(|x_n|^2 + \frac{1}{2} \sum_{i=n+1}^{\infty} |x_i|^2 \right)^{1/2}.$$

Clearly $\| \cdot \|$ is equivalent to the usual l^2 norm. Also, it is easy to see that

$$\|x\|^2 \geqslant \|S_n x\|^2 + \tfrac{1}{2}\|R_n x\|^2, \qquad n = 1, 2, \ldots,$$

and this shows that X satisfies the Gossez–Lami Dozo condition; thus X is $\varDelta$-uniformly convex. However it is easy to see that for any $x \in X$,

$$\limsup_{n \to \infty} \|x - e_n\| \geqslant 1,$$

while for all k, n with $k < n$,

$$\|(1/2^{1/2}) e_k - e_n\| = 1.$$

Thus we conclude

$$A(X, \{e_n\}) \supset (1/2^{1/2}) \,\overline{\mathrm{conv}}\{e_n\}$$

and in particular, $A(X, \{e_n\})$ is not compact.

Spaces or sets in which asymptotic centers are compact have not been completely characterized, but partial results are known. T. C. Lim (1980) has shown that bounded sequences in l^1 have compact asymptotic centers relative to weak*, compact, convex sets. Also, S. Prus (unpublished) has proved that compactness of asymptotic centers in reflexive spaces follows from a stronger version of the Gossez–Lami Dozo condition, one for example satisfied by Day's space D_1, D_∞.

While asymptotic center techniques are very useful theoretically, there are practical disadvantages. For a given sequence $\{x_n\}$ and set K in X, it is usually difficult to describe $A(K, \{x_n\})$ completely. Our next example illustrates this fact.

Example 9.4(a) Consider $X = l^1$ and let $\{x^n\} \subset l^1$ be a sequence with $w^*\text{-}\lim_{n\to\infty} x^n = x$. Thus (thinking of $l^1 = c_0^*$) $\{x^n\}$ is coordinate-wise convergent. It was shown in Chapter 3 that for any $y \in l^1$,

$$r(y, \{x^n\}) = r(x, \{x^n\}) + \|x - y\|. \tag{9.8}$$

This of course implies that $r(x, \{x^n\})$ is the asymptotic radius of $\{x^n\}$ in l^1 and $A(l^1, \{x^n\}) = \{x\}$. Thus the asymptotic center of a weak*-convergent sequence in l^1 with respect to the whole space is precisely its weak*-limit. Suppose now that K is a closed and convex subset of l^1. In view of (9.8) $A(K, \{x^n\})$ consists of all points in K which are nearest x. This set of course may either be empty (especially if K is not weak* compact) or consist of many points.

Example 9.4(b) The situation in the l^p spaces for $1 < p < \infty$ is analogous to the above. The analogue of (9.8) for weakly convergent sequences has the form:

$$(r(y, \{x^n\}))^p = (r(x, \{x^n\}))^p + \|x - y\|^p. \tag{9.9}$$

Thus the asymptotic center of a weakly convergent sequence in l^p coincides with its weak limit, and for any closed and convex subset K of l^p, $A(K, \{x^n\})$ consists of the unique point in K which is nearest x.

The set of points $P_K(x)$ in a given subset K of a Banach space X which are nearest a point $x \in X$ is called the *nearest point projection (or metric projection)* of x on K. Specifically,

$$P_K(x) = \{z \in K : \|x - z\| = \text{dist}(x, K)\}.$$

With this notation the above example shows that

$$A(K, \{x^n\}) = P_K\left(\underset{n\to\infty}{w\text{-}\lim}\, x^n\right)$$

for all weakly convergent sequences $\{x^n\}$ in l^p, $1 < p < \infty$. Correspondingly,

$$A(K, \{x^n\}) = P_K\left(w^*\text{-}\lim_{n \to \infty} x^n\right)$$

for weak* convergent sequences $\{x^n\}$ in l^1. In view of this, if K is a weak* compact, convex subset of l^1 [or a bounded, closed and convex subset of l^p $(1 < p < \infty)$] and if $T: K \to K$ is nonexpansive, then weak* [or weak] subsequential limits of approximate fixed point sequences are always fixed points of T.

If K is a subset of l^1 which is not weak* compact the situation is quite unpredictable.

Example 9.5 We modify Example 3.5. Consider a bounded sequence $\{a_i\}$ of nonnegative numbers and, with $\{e^i\}$ denoting the standard basis in l^1, form the sequence $\{f^i\} \equiv \{(1 + a_i) e^i\}$. Set

$$K = \text{conv}\{f^i\} \equiv \left\{x = \sum_{i=1}^{\infty} \lambda_i f^i \colon \lambda_i \geqslant 0, \sum_{i=1}^{\infty} \lambda_i = 1\right\},$$

and let $\bar{K}^{w^*}$ denote the weak* closure of $\text{conv}(K \cup \{0\})$. Thus

$$\bar{K}^{w^*} = \left\{x = \sum_{i=1}^{\infty} \mu_i f^i \colon \mu_i \geqslant 0, \sum_{i=1}^{\infty} \mu_i \leqslant 1\right\}.$$

Given $x \in \bar{K}^{w^*}$, set $\delta_x = 1 - \sum_{i=1}^{\infty} \mu_i$. Finally, let $a = \inf a_i$ and $N_0 = \{i: a_i = a\}$. It is easy to see that the following properties hold. For any $x \in {}^{w^*}\bar{K}$:

(a) $\text{dist}(x, K) = \delta_x(1 + a)$.
(b) $P_K(x) = \text{conv}\{x + \delta_x f^i \colon i \in N_0\}$.
(c) $P_K(x) = \emptyset$ if $N_0 = \emptyset$; $P_K(x)$ is compact (actually finite dimensional) if N_0 is finite; $P_K(x)$ is nonempty and noncompact if N_0 is infinite.

Next we observe that K has the fixed point property (f.p.p.) if and only if N_0 is nonempty but finite. To see this, note that if N_0 has this property and if $T: K \to K$ is nonexpansive, then any weak* convergent, approximate fixed point sequence has a finite dimensional (T-invariant) asymptotic center. On the other hand, if N_0 is empty then there exists a strictly decreasing subsequence $\{a_{i_k}\}$ of $\{a_i\}$ such that $\lim_{k \to \infty} a_{i_k} = a$. Let $N_k = \{i: a_{i_k} < a_i < a_{i_{k-1}}\}$, $k = 1, 2, \ldots$, $(a_{i_0} = +\infty)$. Define $T: K \to K$ by

$$T: x = \sum_{i=1}^{\infty} \lambda_i f^i \to \sum_{k=1}^{\infty} \mu_k f^{i_k+1},$$

where

$$\mu_k = \sum_{i \in N_k} \lambda_i.$$

Then T is nonexpansive and fixed point free. An example of a fixed point free, nonexpansive mapping may be constructed in a similar way in the case N_0 is infinite (see Goebel and Kuczumow, 1978a).

The above example shows that the f.p.p. may be very 'unstable'. By altering the choice of $\{a_i\}$ and modifying the above construction it is possible to verify the following facts (see Goebel and Kuczumow, 1978a).

(a)   The f.p.p. is not hereditary, in the sense that a closed, convex subset K_1 of K may fail to have the f.p.p. even if K itself has that property.

(b)   The f.p.p. is not preserved in passing to limits, i.e., there exists a sequence $\{K_n\}$ of sets having the f.p.p. whose limit K_∞, in the Hausdorff metric sense, fails to have the f.p.p.

(c)   There exists a descending sequence $\{K_n\}$ of closed convex sets with the property that for each n, K_{2n+1} has the f.p.p. while K_{2n} fails to have it. Moreover, this sequence may be constructed in such a fashion that the intersection $K_\infty = \bigcap K_n$ may or may not have the f.p.p. It is even possible to accomplish this while assuring $K_\infty = \lim_{n \to \infty} K_n$.

The above facts clearly show that the f.p.p. for nonexpansive mappings is a delicate concept which may depend on the nature of the set K rather than on the nature of space in which it resides.

Our next example illustrates another application of the asymptotic center approach.

Example 9.6   Let X_1 be an arbitrary uniformly convex Banach space, and consider the product space $X = X_1 \times l^1$ with the norm

$$\|(x, y)\| = \max\{\|x\|_{X_1}, \|y\|_1\}, \qquad x \in X_1, y \in l^1.$$

Let K_1 be a closed, bounded subset of X_1, let B denote the unit ball in l^1, and set $K = K_1 \times B$. Since X is nonreflexive K fails to be weakly compact. However we claim that K has the f.p.p. To see this, first note that for any sequence $\{(x_n, y_n)\}$ in K, $\{y_n\}$ may be assumed, by passing to a subsequence, to be weak* convergent, say to $y \in B$. Also, the asymptotic center of $\{x_n\}$ in K_1 consists of exactly one point. If r_1 and r_2 denote the respective asymptotic radii of $\{x_n\}$ and $\{y_n\}$, then

$$r(K, \{(x_n, y_n)\}) = \max\{r_1, r_2\}.$$

If $\{x\} = A(K_1, \{x_n\})$ and $r_1 = r_2$, then

$$A(K, \{(x_n, y_n)\}) = \{(x, y)\}.$$

If $r_1 < r_2$, then

$$A(K, \{(x_n, y_n)\}) = \{z \in K_1 : \limsup_{n \to \infty} \|z - x_n\| \leqslant r_2\} \times \{y\}$$

$$= A_{r_2 - r_1}(K_1, \{x_n\}) \times \{y\}. \tag{9.10}$$

Since this set is isometric to a closed, convex subset of K_1, it behaves like a subset of a uniformly convex space.

Finally, if $r_1 > r_2$, then

$$A(K, \{(x_n, y_n)\}) = \{x\} \times \{z \in B : \limsup_{n \to \infty} \|z - y_n\|_1 \leqslant r_1\}$$

$$= \{x\} \times A_{r_1 - r_2}(B, \{y_n\})$$

$$= \{x\} \times \{B \cap B(y; r_1 - r_2)\}.$$

This set may be viewed (isometrically) as a weak* compact subset of l^1.

Now let $T: K \to K$ be nonexpansive. In view of the above, the following procedure may be used to obtain fixed points of T. Let $\{(x_n, y_n)\}$ be an approximate fixed point sequence for T. In the above notation, if $r_1 = r_2$ then $T(x, y) = (x, y)$. If $r_1 < r_2$, select $(u, y) \in A(K, \{(x_n, y_n)\})$ and note that T must have a fixed point lying in the asymptotic center of $\{T^n(u, y)\}$ with respect to the set (9.10). If $r_1 > r_2$, then again T has a fixed point in $A(K, \{(x_n, y_n)\})$.

Many of the examples in this chapter have dealt with either iterative sequences or approximate fixed point sequences. Since, in general, an iterative sequence need not be an approximate fixed point sequence, the following definition is a natural one.

Definition 9.1 A nonexpansive mapping $T: K \to K$ is said to be *asymptotically regular* if for any $x \in K$,

$$\lim_{n \to \infty} \|T^n x - T^{n+1} x\| = 0.$$

The following observation is due to J. Alexander (1985, personal communication).

Example 9.7 Let $T: [0, 1] \to [0, 1]$ be nonexpansive and let $S = \frac{1}{2}(I + T)$.

Then S is also nonexpansive. Assume that for some integer n and $x \in [0, 1]$, $|S^{n-2}x - S^{n-1}x| > 1/n$. Then $|x - Sx| \geqslant |Sx - S^2 x| \geqslant \cdots \geqslant |S^{n-2}x - S^{n-1}x| > 1/n$, and also

$$|S^{n-1}x - TS^{n-2}x| = |S^{n-2}x - S^{n-1}x| > 1/n.$$

It is easy to see that the iteration

$$\{x, Sx, S^2 x, \ldots, S^{n-1}x, TS^{n-2}x\}$$

is monotone, so the above inequalities yield the contradiction $|x - TS^{n-2}x| > 1$.

We conclude from the above that if $\epsilon > 0$ and if $x \in [0, 1]$, then for any nonexpansive mapping $T: [0, 1] \to [0, 1], |S^n x - S^{n+1}x| \leqslant \epsilon$ whenever $n \geqslant \epsilon^{-1} - 2$. This fact is a very special case of the following qualitative result.

Theorem 9.4 *Let X be an arbitrary Banach space, let K be a bounded, closed and convex subset of X, and let $\mathscr{F}$ denote the collection of all nonexpansive self-mappings of K. Fix $\alpha \in (0, 1)$ and for $T \in \mathscr{F}$, set $T_\alpha = (1 - \alpha)I + \alpha T$. Then T_α is asymptotically regular on K. Moreover, the sequence $\{\|T_\alpha^n x - T_\alpha^{n+1} x\|\}$ converges to 0 uniformly for $x \in K$ and $T \in \mathscr{F}$. Precisely, if $\epsilon > 0$ then there exists an integer N, depending only on ϵ (and K) such that if $n \geqslant N$, if $x \in K$, and if $T \in \mathscr{F}$, then $\|T_\alpha^n x - T_\alpha^{n+1} x\| \leqslant \epsilon$.*

Before proving the above theorem we observe that for each $T \in \mathscr{F}$ and $\alpha \in (0, 1)$ the fixed point sets of T and T_α coincide. Thus the iterative sequences $\{T_\alpha^n x\}$ for $x \in K$ form approximate fixed point sequences for T_α and their asymptotic centers are T_α invariant. This fact can simplify efforts to obtain fixed points of T.

We base the proof of Theorem 9.4 on the following technical lemma, which establishes the convergence of $\{\|T_\alpha^n x - T_\alpha^{n+1} x\|\}$ to zero for any fixed $x \in K$ and $T \in \mathscr{F}$ separately, a result first proved by Ishikawa (1976). The proof is completed by showing the required uniformity.

Lemma 9.4 *Let X be a Banach space, K a bounded, closed and convex subset of X, and $T: K \to K$ nonexpansive. For $\alpha \in (0, 1)$, set $T_\alpha = (1 - \alpha)I + \alpha T$, fix $x_0 \in K$, and define the sequences $\{x_n\}$ and $\{y_n\}$ as follows:*

$$x_{n+1} = T_\alpha x_n; \qquad y_n = Tx_n, \qquad n = 0, 1, 2, \ldots.$$

Then for each $i, n \in \mathbb{N}$,

$$\|y_{i+n} - x_i\| \geqslant (1 - \alpha)^{-n}[\|y_{i+n} - x_{i+n}\| - \|y_i - x_i\|] + (1 + n\alpha)\|y_i - x_i\|, \quad (9.12)$$

and

$$\lim_{n \to \infty} \|x_n - Tx_n\| = 0. \qquad (9.13)$$

Proof of (9.12) We proceed by induction on n, assuming that (9.12) holds for a given n and all i. ((9.12) is trivial if $n=0$.) Replacing i with $i+1$ in (9.12) yields:

$$\|y_{i+n+1}-x_{i+1}\| \geqslant (1-\alpha)^{-n}[\|y_{i+n+1}-x_{i+n+1}\|-\|y_{i+1}-x_{i+1}\|]$$
$$+(1+n\alpha)\|y_{i+1}-x_{i+1}\|. \tag{9.14}$$

Also

$$\|y_{i+n+1}-x_{i+1}\| \leqslant (1-\alpha)\|y_{i+n+1}-x_i\|+\alpha\|y_{i+n+1}-y_i\|$$

$$\leqslant (1-\alpha)\|y_{i+n+1}-x_i\|+\alpha\sum_{k=0}^{n}\|y_{i+k+1}-y_{i+k}\|$$

$$\leqslant (1-\alpha)\|y_{i+n+1}-x_i\|+\alpha\sum_{k=0}^{n}\|x_{i+k+1}-x_{i+k}\|. \tag{9.15}$$

Combining (9.14) and (9.15):

$$\|y_{i+n+1}-x_i\| \geqslant (1-\alpha)^{-(n+1)}[\|y_{i+n+1}-x_{i+n+1}\|-\|y_{i+1}-x_{i+1}\|]$$
$$+(1-\alpha)^{-1}(1+n\alpha)\|y_{i+1}-x_{i+1}\|$$
$$-\alpha(1-\alpha)^{-1}\sum_{k=0}^{n}\|x_{k+i+1}-x_{k+i}\|.$$

Since $\|x_{k+i+1}-x_{k+i}\| \equiv \alpha\|y_{k+i}-x_{k+i}\|$, and since the sequence $\{\|y_n-x_n\|\}$ is decreasing and $1+n\alpha \leqslant (1-\alpha)^{-n}$, we have:

$$\|y_{i+n+1}-x_i\| \geqslant (1-\alpha)^{-(n+1)}[\|y_{i+n+1}-x_{i+n+1}\|-\|y_{i+1}-x_{i+1}\|]$$
$$+(1-\alpha)^{-1}(1+n\alpha)\|y_{i+1}-x_{i+1}\|$$
$$-\alpha^2(1-\alpha)^{-1}(n+1)\|y_i-x_i\|$$
$$=(1-\alpha)^{-(n+1)}[\|y_{i+n+1}-x_{i+n+1}\|-\|y_i-x_i\|]$$
$$+[(1-\alpha)^{-1}(1+n\alpha)-(1-\alpha)^{-(n+1)}]\|y_{i+1}-x_{i+1}\|$$
$$+[(1-\alpha)^{-(n+1)}-\alpha^2(1-\alpha)^{-1}(n+1)]\|y_i-x_i\|$$
$$\geqslant (1-\alpha)^{-(n+1)}[\|y_{i+n+1}-x_{i+n+1}\|-\|y_i-x_i\|]$$
$$+[(1-\alpha)^{-1}(1+n\alpha)-(1-\alpha)^{-(n+1)}]\|y_i-x_i\|$$
$$+[(1-\alpha)^{-(n+1)}-\alpha^2(1-\alpha)^{-1}(n+1)]\|y_i-x_i\|$$
$$=(1-\alpha)^{-(n+1)}[\|y_{i+n+1}-x_{i+n+1}\|-\|y_i-x_i\|]$$
$$+(1+(n+1)\alpha)\|y_i-x_i\|.$$

Thus (9.12) holds for $n+1$, completing the proof.

Proof of (9.13) Let $d = \operatorname{diam} K$, and suppose $\lim_{n \to \infty} \|y_n - x_n\| = r > 0$. Select $N \geqslant d/r\alpha$ and let $\epsilon > 0$ satisfy $\epsilon(1 - \alpha)^{-N} < r$. Since the sequence $\{\|y_n - x_n\|\}$ is decreasing, there exists an integer i such that

$$0 \leqslant \|y_i - x_i\| - \|y_{i+N} - x_{i+N}\| \leqslant \epsilon.$$

Therefore, using (9.12), we obtain the contradiction:

$$d + r \leqslant (1 + N\alpha)r \leqslant (1 + N\alpha)\|y_i - x_i\|$$
$$\leqslant \|y_{i+N} - x_i\| + (1 + \alpha)^{-N}\epsilon$$
$$< d + r.$$

Consequently $r = 0$ and the proof is complete.

Proof of Theorem 9.4 With X, K and $\mathscr{F}$ as in the theorem consider the Banach space

$$B = B(K, X)$$

of all bounded mappings $\varphi : K \to X$ with the standard uniform norm:

$$\|\varphi\|_B = \sup\{\|\varphi x\|_X : x \in K\}.$$

Let $D \subset B$ denote the closed, bounded and convex subset of B consisting of all self-mappings of K $(D = K^K)$ and let

$$\mathscr{X} = B(\mathscr{F}, B(K, X)),$$

i.e., $\mathscr{X}$ is the space of all bounded functions $f : \mathscr{F} \to B(K, X) = B$ with the uniform norm

$$\|f\| = \sup\{\|f(T)\|_B : T \in \mathscr{F}\}.$$

The set

$$\mathscr{K} = \{f \in \mathscr{X} : f(T) \in D \text{ for all } T \in \mathscr{F}\}$$

is obviously bounded, closed and convex in $\mathscr{X}$.

Now define $F : \mathscr{K} \to \mathscr{K}$ as follows: For $f \in \mathscr{X}$, set

$$F(f)(T) = T \circ f(T), \quad T \in \mathscr{F},$$

and observe that for each $f, g \in \mathscr{K}$,

$$\|F(f) - F(g)\| = \sup\{\|T \circ f(T)(x) - T \circ g(T)(x)\|_X : T \in \mathscr{F}, x \in K\}$$
$$\leqslant \sup\{\|f(T)(x) - g(T)(x)\|_X : T \in \mathscr{F}, x \in K\}$$
$$= \|f - g\|.$$

Therefore F is nonexpansive and, in view of Lemma 9.4, if $F_\alpha = (1-\alpha)I + \alpha F$, $\alpha \in (0, 1)$, then for any $f \in \mathcal{K}$

$$\lim_{n \to \infty} \|F_\alpha^n f - F_\alpha^{n+1} f\| = 0$$

The conclusion of Theorem 9.4 now follows upon applying the above to $f \in \mathcal{K}$ defined by $f(T) = id_K$ for all $T \in \mathcal{F}$.

Theorem 9.4 has the following immediate corollary.

Corollary 9.1 *Let X be a Banach space, K a closed, convex subset of X, and $T: K \to K$ a nonexpansive mapping for which $T(K)$ is compact. Then for each $\alpha \in (0, 1)$ the iterates of the mapping $T_\alpha = (1-\alpha)I + \alpha T$ converge to a fixed point of T.*

The proof of Theorem 9.4 establishes the *existence* of an integer N such that if $T: K \to K$ is nonexpansive, if $x_0 \in K$, and if $n \geqslant N$, then

$$\|S^n(x_0) - S^{n+1}(x_0)\| \leqslant \epsilon$$

where $S = \frac{1}{2}(I + T)$. On the other hand, a precise estimate on the optimal magnitude of such an N is given in Example 9.6 in the case $K = [0, 1]$. While it is unlikely that a common fixed value of N exists for all sets K of a given diameter, if the sets K all lie in the class of uniformly convex Banach spaces whose moduli of convexity are all bounded below by some fixed positive modulus, then a common value for N does exist. This fact is a consequence of the following (Kirk and Martinez Yanez, 1990):

Theorem 9.5 *Let X be a uniformly convex Banach space with modulus of convexity δ, let K be a nonempty, bounded, closed and convex subset of X with diam $K = d$, and let $\epsilon > 0$ $(\epsilon \leqslant d/2)$. If $T: K \to K$ is nonexpansive and if $S = \frac{1}{2}(I + T)$, then for any $x \in K$, $\|S^n x - S^{n+1} x\| \leqslant \epsilon$ for all n satisfying*

$$(1 - \delta(2\epsilon/d))^n \leqslant \epsilon/d. \tag{9.16}$$

Proof Let $x \in K$, and let $z \in K$ be a fixed point of S. Since $\{\|S^n x - S^{n+1} x\|\}$ is monotone decreasing, if $\|S^k x - S^{k+1} x\| \leqslant \epsilon$ holds for some $k \leqslant n$ there is nothing to prove. So we assume

$$\epsilon < \|S^n x - S^{n+1} x\| \leqslant \|S^{n-1} x - S^n x\| \leqslant \cdots \leqslant \|x - Sx\| \leqslant d. \tag{9.17}$$

Since $\|S^{n-1} x - S^n x\| > \epsilon$, $\|S^{n-1} x - TS^{n-1} x\| > 2\epsilon$. Also, $\|TS^{n-1} x - z\| \leqslant \|S^{n-1} x - z\|$ and therefore, since $S^n x = \frac{1}{2}(S^{n-1} x + TS^{n-1} x)$,

$$\|S^n x - z\| \leqslant \left(1 - \delta\left(\frac{2\epsilon}{\|S^{n-1} x - p\|}\right)\right)\|S^{n-1} x - z\|. \tag{9.18}$$

In view of this,

$$\|S^n x - z\| \le \prod_{j=1}^{n}\left(1 - \delta\left(\frac{2\epsilon}{\|S^{n-j}x - p\|}\right)\right)\|x - z\|$$

and by monotocity of δ

$$\|S^n x - z\| \le \left(1 - \delta\left(\frac{2\epsilon}{d}\right)\right)^n d$$

$$\le (\epsilon/d)d$$

$$= \epsilon.$$

Since z is a fixed point of T with T nonexpansive, $\|TS^n x - z\| \le \epsilon$; hence $\|S^n x - TS^n x\| \le 2\epsilon$ yielding

$$\|S^n x - S^{n+1} x\| = \tfrac{1}{2}\|S^n x - TS^n x\| \le \epsilon.$$

Example 9.8 In cases where the modulus of convexity is explicitly known the estimate (9.16) can be improved. If $X = l^p, 2 \le p < \infty$, then $\delta(\epsilon) = 1 - (1 - (\epsilon/2)^p)^{1/p}$. Computing directly from (9.18)

$$\|Sx - z\| \le (1 - \delta(2\epsilon/d))d$$

$$= (1 - (\epsilon/d)^p)^{1/p}d$$

$$= (d^p - \epsilon^p)^{1/p};$$

$$\|S^2 x - z\| \le \left(1 - \delta\left(\frac{2\epsilon}{(d^p - \epsilon^p)^{1/p}}\right)\right)(d^p - \epsilon^p)^{1/p}$$

$$= (d^p - 2\epsilon^p)^{1/p};$$

$$\vdots$$

$$\|S^n x - z\| \le (d^p - n\epsilon^p)^{1/p}d, \qquad n = 1, 2, \ldots.$$

Also, $(d^p - n\epsilon^p)^{1/p} \le \epsilon$ if and only if $n \ge (d/\epsilon)^p - 1$. This provides a smaller value for n than that given by (9.16).

Remarks Theorem 9.4 actually holds with $\{T_\alpha^n x\}$ replaced by the more general iteration scheme $\{x_n\}$ defined as follows. Let $\{\alpha_n\} \subset [0, 1)$ be a sequence for which $0 \le \alpha_n \le b < 1$ and $\Sigma_{n=0}^{\infty}\alpha_n = +\infty$; for $T \in \mathscr{F}$ and $x_0 \in K$ define

$$x_{n+1} = (1 - \alpha_n)x_n + \alpha_n Tx_n, \qquad n = 0, 1, 2, \ldots.$$

In the general case the details may be found in (Goebel and Kirk, 1983); the earlier 'nonuniform' version is due to Ishikawa (1976). Edelstein and O'Brien (1978) give a proof of Theorem 9.4 for a single mapping.

Corollary 9.1 has an earlier history. In 1955 Krasnoselskii (1955) obtained this result for X uniformly convex (and $\alpha_n \equiv \frac{1}{2}$), and subsequently Edelstein (1966) observed that it remained true for strictly convex X. Since any restriction on X eventually proved to be unnecessary, it is natural to ask whether the compactness assumption might be exchanged for nicer behavior of X. The answer is no. In (1975) Genel and Lindenstrauss constructed an example of a bounded, closed and convex set K in the Hilbert space l^2 and a nonexpansive mapping $T: K \to K$ with the property that the iterates of $\{T_{1/2}x\}$ fail to converge to a fixed point of T for some $x \in K$.

Finally we remark that Baillon, Bruck, and Reich (1978) have extended Corollary 9.1 in another direction. They show that if $T: K \to K$ is non-expansive for an arbitrary convex set K and if $T_\alpha = (1 - \alpha)I + \alpha T$, $\alpha \in (0, 1)$, then T_α always satisfies

$$\lim_{n \to \infty} \| T_\alpha^{n+1}x - T_\alpha^n x \| = (1/k) \lim_{n \to \infty} \| T_\alpha^{n+k}x - T_\alpha^n x \| = \lim_{n \to \infty} (1/n)\| T_\alpha^n x \|, x \in K.$$

10

Weak sequential approximations

In view of the fact that Theorem 4.1 involves the weak as well as the norm topology, one might expect that in certain circumstances the weak topology could be utilized to approximate fixed points of nonexpansive mappings. Indeed, such a scheme exists (and is implicit in Göhde's original proof (1965)). However we begin this chapter with another approach, one suggested by the observations of Example 9.3. There it is seen that in the l^p spaces, $1 \leqslant p < \infty$, the asymptotic center of a sequence coincides with its weak (weak* in the case of l^1) limit. An example due to Z. Opial illustrates the fact that this is far from true in general.

Example 10.1 Consider the space $L^p[0, 2\pi]$ for $1 < p < \infty$ and let θ be a periodic real-valued function of period 2π such that

$$\theta(t) = \begin{cases} 1 & \text{for } 0 \leqslant t \leqslant (\tfrac{4}{3})\pi \\ -2 & \text{for } (\tfrac{4}{3})\pi < t \leqslant 2\pi. \end{cases}$$

Generate the sequence of functions $\{\theta_n\}$ by setting $\theta_n(t) = \theta(nt)$ and observe that for any step function ψ,

$$\lim_{n \to \infty} \int_0^{2\pi} \theta_n(t)\psi(t)\,dt = 0.$$

Since the step functions form a dense subset of $L^q[0, 2\pi] = (L^p[0, 2\pi])^*$, it follows that w-$\lim_{n \to \infty} \theta_n = 0$. However, for any constant function c close to zero,

$$\|\theta_n - c\|^p = \int_0^{2\pi} |\theta(nt) - c|^p\,dt$$

$$= (1/n) \int_0^{2n\pi} |\theta(t) - c|^p\,dt$$

$$= \int_0^{2\pi} |\theta(t) - c|^p\,dt$$

$$= \tfrac{4}{3}\pi(1 - c)^p + \tfrac{2}{3}\pi(2 + c)^p.$$

104

Thus

$$r^p(c, \{\theta_n\}) = \tfrac{2}{3}\pi[2(1-c)^p + (2+c)^p]$$

and

$$\frac{\mathrm{d}}{\mathrm{d}c}[r^p(c, \{\theta_n\})] = \tfrac{2}{3}\pi p[(2+c)^{p-1} - 2(1-c)^{p-1}];$$

in particular

$$\frac{\mathrm{d}}{\mathrm{d}c}[r^p(0, \{\theta_n\})] = \tfrac{2}{3}\pi p(2^{p-1} - 2),$$

which is different from zero for $p \neq 2$. This shows that $r(c, \{\theta_n\})$ does not attain its minimum for $c = 0$ unless $p = 2$. Therefore asymptotic centers and weak limits do not always coincide.

In 1967a, Opial introduced a class of spaces which includes the l^p spaces, $1 < p < \infty$, for which weak limits and asymptotic centers of sequences coincide. To discuss this class we need some further definitions:

For a point x in a (real) Banach space X and a functional $x^* \in X^*$, as usual we use the pairing $\langle x, x^* \rangle$ to denote $x^*(x)$. Suppose μ is a continuous strictly increasing real-valued function on $\mathbb{R}^+$ satisfying $\mu(0) = 0$ and $\lim_{t \to \infty} \mu(t) = +\infty$.

Definition 10.1 A mapping $J : X \to X^*$ is called a *duality mapping* with gauge function μ if for every $x \in X$,

$$\langle x, Jx \rangle = \|Jx\| \|x\| = \mu(\|x\|)\|x\|.$$

Such a mapping J is said to be *weakly sequentially continuous* if J is sequentially continuous relative to the weak topologies on both X and X^*.

Example 10.2 The spaces $l^p, 1 < p < \infty$, possess duality mappings which are weakly sequentially continuous. To see this, let $(x_1, x_2, \ldots) \in l^p$, set $Jx = (\operatorname{sgn} x_1 |x_1|^{p-1}, \operatorname{sgn} x_2 |x_2|^{p-1}, \ldots)$, and let $\mu(r) = r^{p/q}$ where $p^{-1} + q^{-1} = 1$. Since

$$\left(\sum_{i=1}^{\infty} |x_i|^{(p-1)q} \right)^{1/q} = \left(\sum_{i=1}^{\infty} |x_i|^p \right)^{1/q},$$

$Jx \in l^q$. Also $\mu(\|x\|) = \|x\|^{p/q} = \|Jx\|$, and

$$\langle x, Jx \rangle = \sum_{i=1}^{\infty} x_i \operatorname{sgn} x_i |x_i|^{p-1} = \sum_{i=1}^{\infty} |x_i|^p = \|x\|^p$$

$$= \|x\| \cdot \|Jx\|.$$

Thus J is a duality mapping. To see that J is weakly sequentially continuous one need only go through routine calculations, keeping in mind that in this setting weak convergence coincides with boundedness and coordinatewise convergence.

Theorem 10.1 (Browder, 1967; Opial, 1967a) *Let X be a Banach space with weakly sequentially continuous duality mapping J, and suppose $\{x_n\}$ converges weakly to x_0 in X. Then for any $x \in X$,*

$$\liminf_{n \to \infty} \| x_n - x_0 \| \leqslant \liminf_{n \to \infty} \| x_n - z \|. \tag{10.1}$$

Moreover if X is uniformly convex, equality holds if and only if $x_0 = z$.

Proof (cf., Belluce 1967–68) Let μ be the gauge function for J and set $\varphi(r) = \int_0^r \mu(t)\,dt$. For $x \in X$, set $g(x) = \varphi(\|x\|)$. Observe that for $x, y \in X$,

$$\langle x - y, Jy \rangle = \langle x, Jy \rangle - \langle y, Jy \rangle \leqslant \| Jy \| \| x \| - \| Jy \| \| y \|$$
$$= \mu(\| y \|)(\| x \| - \| y \|).$$

Also,

$$g(x) - g(y) = \int_0^{\|x\|} \mu(t)\,dt - \int_0^{\|y\|} \mu(t)\,dt.$$

Thus if $\| y \| < \| x \|$, then $g(x) - g(y) \geqslant \mu(\| y \|)(\| x \| - \| y \|)$, and if $\| x \| < \| y \|$, then $g(y) - g(x) \leqslant \mu(\| y \|)(\| y \| - \| x \|)$. So in either case $g(x) - g(y) \geqslant \mu(\| y \|)(\| x \| - \| y \|) \geqslant \langle x - y, Jy \rangle$. In particular,

$$\varphi(\| x_n - x \|) - \varphi(\| x_n - x_0 \|) \geqslant \langle x_0 - x, J(x_n - x_0) \rangle.$$

But $\underset{n \to \infty}{\text{w-lim}}\ x_n = x_0$ so $\underset{n \to \infty}{\text{w-lim}}\ J(x_n - x_0) = 0$. Therefore

$$\liminf_{n \to \infty} [\varphi(\| x_n - x \|) - \varphi(\| x_n - x_0 \|)] \geqslant 0.$$

Now select a subsequence $\{x_{n_k}\}$ of $\{x_n\}$ for which $\lim_{k \to \infty} \| x_{n_k} - x \| = \liminf_{n \to \infty} \| x_n - x \|$. Then $\liminf_{n \to \infty} \| x_n - x_0 \| \leqslant \liminf_{k \to \infty} \| x_{n_k} - x_0 \|$ and moreover, $\varphi(\liminf_{k \to \infty} \| x_{n_k} - x_0 \|) = \liminf_{k \to \infty} \varphi(\| x_{n_k} - x_0 \|)$. Thus

$$0 \leqslant \liminf_{k \to \infty} [\varphi(\| x_{n_k} - x \|) - \varphi(\| x_{n_k} - x_0 \|)]$$
$$= \lim_{k \to \infty} \varphi(\| x_{n_k} - x \|) - \limsup_{k \to \infty} \varphi(\| x_{n_k} - x_0 \|).$$

We now have

$$\liminf_{n \to \infty} \varphi(\|x_n - x_0\|) \leqslant \liminf_{k \to \infty} \varphi(\|x_{n_k} - x_0\|)$$
$$\leqslant \limsup_{k \to \infty} \varphi(\|x_{n_k} - x_0\|)$$
$$\leqslant \liminf_{k \to \infty} \varphi(\|x_{n_k} - x\|)$$
$$= \liminf_{n \to \infty} \varphi(\|x_n - x\|).$$

Since φ is strictly increasing, the first part of the theorem is proved.

Now assume X is uniformly convex and suppose

$$c = \liminf_{n \to \infty} \|x_n - x_0\| = \liminf_{n \to \infty} \|x_n - x\|.$$

Select a subsequence $\{x_{n_k}\}$ of $\{x_n\}$ as above. Then $c \leqslant \liminf_{k \to \infty} \|x_{n_k} - x_0\| \leqslant \limsup_{k \to \infty} \|x_{n_k} - x_0\| \leqslant \liminf_{k \to \infty} \|x_{n_k} - x\| = c$. Therefore

$$\lim_{k \to \infty} \|x_{n_k} - x_0\| = c = \lim_{k \to \infty} \|x_{n_k} - x\|.$$

Now suppose $x \neq x_0$ and let $y = \frac{1}{2}(x + x_0)$. Then

$$\liminf_{k \to \infty} \|y - x_{n_k}\| \leqslant [1 - \delta[\|x - x_0\|/c]]c < c.$$

However by the first part of the theorem,

$$c = \liminf_{n \to \infty} \|x_n - x_0\| \leqslant \liminf_{n \to \infty} \|x_n - y\| \leqslant \liminf_{k \to \infty} \|x_{n_k} - y\|,$$

so if $x \neq x_0$ we have a contradiction.

Remark Gossez and Lami Dozo (1972) have shown that the last conclusion of the above theorem holds if X is merely assumed to be reflexive. Such spaces are said to satisfy *Opial's condition*. Specifically:

Definition 10.2 A Banach space X is said to satisfy *Opial's condition* if whenever a sequence $\{x_n\}$ in X converges weakly to x_0, then for $x \neq x_0$,

$$\liminf_{n \to \infty} \|x_n - x_0\| < \liminf_{n \to \infty} \|x_n - x\|. \tag{10.2}$$

Example 10.1 and Theorem 10.1 show, respectively, that the L^p space, $p \neq 2$, do not satisfy Opial's condition while all the l^p spaces do $(1 < p < \infty)$. Thus Opial's condition is independent of uniform convexity. On the other hand, Gossez and Lami Dozo (1972) have observed that all such spaces have normal structure.

Theorem 10.2 *If X is a reflexive Banach space which satisfies Opial's condition, then X has normal structure.*

Proof If X fails to have normal structure, then X contains a diametral sequence $\{x_n\}$ which we may assume converges weakly to 0. Since $\{x_n\}$ is diametral,

$$\lim_{n \to \infty} \text{dist}(x_{n+1}, \text{conv}\{x_1, x_2, \ldots, x_n\}) = \text{diam}\{x_n\}.$$

In particular, if $y \in \text{conv}\{x_1, x_2, \ldots\}$, then $\lim_{n \to \infty} \| y - x_n \| = \text{diam}\{x_n\}$, and the same holds if $y \in \overline{\text{conv}}\{x_1, x_2, \ldots\}$. Therefore, taking $y = 0$, $\lim_{n \to \infty} \| x_n \| = \text{diam}\{x_n\}$. Since also $\lim_{n \to \infty} \| x_1 - x_n \| = \text{diam}\{x_n\}$, this contradicts (10.2).

Spaces which satisfy Opial's condition have another nice property related to fixed point theory. A mapping f defined on a subset D of a Banach space X (and taking values in X) is said to be *demiclosed* if for any sequence $\{u_j\}$ in D the following implication holds:

$$\text{w-lim}_{j \to \infty} u_j = u \quad \text{and} \quad \lim_{j \to \infty} \| f(u_j) - w \| = 0$$

implies

$$u \in D \quad \text{and} \quad f(u) = w.$$

Theorem 10.3 *Let X be a reflexive Banach space which satisfies Opial's condition, let K be a closed, convex subset of X, and suppose $T: K \to X$ is nonexpansive. Then the mapping $f = I - T$ is demiclosed on K.*

Proof Suppose $\{u_n\}$ in K satisfies $\text{w-lim}_{n \to \infty} u_n = u$ and

$$\lim_{n \to \infty} \| u_n - Tu_n - w \| = 0.$$

If the mapping T is replaced with the mapping T_w defined by $T_w x = Tx + w$, then $\lim_{n \to \infty} \| u_n - T_w u_n \| = 0$ and T_w is also nonexpansive. We now have

$$\| T_w u - u_n \| \leqslant \| T_w u - T_w u_n \| + \| T_w u_n - u_n \|;$$

hence

$$\liminf_{n \to \infty} \| T_w u - u_n \| \leqslant \liminf_{n \to \infty} \| u - u_n \|.$$

By Opial's condition, $T_w u = u$; i.e., $u - Tu = w$.

It is perhaps surprising that the conclusion of the above theorem holds in all uniformly convex spaces in spite of the fact that not all such spaces have weakly sequentially continuous duality mappings – a fact noted by Browder in 1968.

Theorem 10.4 (Demiclosedness Principle) *Let X be a uniformly convex Banach space, K a closed and convex subset of X, and $T: K \rightarrow X$ non-expansive. Then the mapping $f = I - T$ is demiclosed on K.*

This theorem is an immediate consequence of the following two propositions.

Proposition 10.1 *Suppose K is a bounded, convex subset of a uniformly convex space X and suppose $T: K \rightarrow X$ is nonexpansive. Then for $\{u_n\}$, $\{v_n\}$ in K and $z_n = \frac{1}{2}(u_n + v_n)$,*

$$\left. \begin{array}{l} \lim_{n \to \infty} \| u_n - Tu_n \| = 0 \\ \lim_{n \to \infty} \| v_n - Tv_n \| = 0 \end{array} \right\} \Rightarrow \lim_{n \to \infty} \| z_n - Tz_n \| = 0.$$

Proof Suppose the implication fails. Then there exist sequences $\{u_n\}$, $\{v_n\}$ in K and $\epsilon > 0$ satisfying $\lim_{n \to \infty} \| u_n - Tu_n \| = 0$ and $\lim_{n \to \infty} \| v_n - Tv_n \| = 0$ while, if $z_n = \frac{1}{2}(u_n + v_n)$, then $\| z_n - Tz_n \| \geqslant \epsilon > 0$. In view of this we may suppose, by passing to a subsequence, that for some $r > 0$, $\lim_{n \to \infty} \| u_n - z_n \| = \lim_{n \to \infty} \| v_n - z_n \| = r$. Let $d = \text{diam } K$ and choose $t > 0$ so that $t < \epsilon / d$. Then clearly $t < \epsilon / \| u_n - z_n \|$ and so for n sufficiently large,

$$t < \epsilon / [\| u_n - Tu_n \| + \| u_n - z_n \|].$$

Also,

$$\| u_n - Tz_n \| \leqslant \| u_n - Tu_n \| + \| Tu_n - Tz_n \|$$
$$\leqslant \| u_n - Tu_n \| + \| u_n - z_n \|.$$

Since all the above inequalities hold if u_n is replaced by v_n, we have (for large n) by (5.3):

$$\| u_n - v_n \| \leqslant [\| u_n - \tfrac{1}{2}(z_n + Tz_n) \| + \| v_n - \tfrac{1}{2}(z_n + Tz_n) \|]$$
$$\leqslant [(\| u_n - Tu_n \| + \| u_n - z_n \|) + (\| v_n - Tv_n \| + \| v_n - z_n \|)](1 - \delta(t)).$$

Letting $n \to \infty$ we obtain the contradiction

$$2r \leqslant 2r(1 - \delta(t)).$$

Proposition 10.2 *Suppose K is a bounded, closed and convex subset of a uniformly convex space X and suppose $T: K \rightarrow X$ is a nonexpansive mapping which satisfies $\inf \{ \| x - Tx \| : x \in K \} = 0$. Then T has a fixed point in K.*

Proof Let R denote the set of numbers $r > 0$ for which $B(0; r) \cap K \neq \emptyset$ and $\inf \{ \| x - Tx \| : x \in B(0; r) \cap K \} = 0$, and let $r_0 = \inf R$. Since K is bounded,

$r_0 < \infty$, and if $r_0 = 0$ then $0 \in K$ and $T0 = 0$. So we may suppose $r_0 > 0$. It is clearly possible to select $x_n \in B(0; r_0 + 1/n) \cap K$ for each n so that $\lim_{n \to \infty} \| x_n - Tx_n \| = 0$. Since any strongly convergent subsequence of $\{x_n\}$ would have a fixed point of T as its limit, we may suppose there exists $\epsilon > 0$ and a subsequence $\{x_{n_k}\}$ of $\{x_n\}$ such that $\| x_{n_k} - x_{n_{k+1}} \| \geqslant \epsilon, k = 1, 2, \ldots$. For each k, let $m_k = \frac{1}{2}(x_{n_k} + x_{n_{k+1}})$. Then if $t > 0$ is any number smaller than $\epsilon / r_0, t \leqslant \epsilon / \| x_k \|$ for k sufficiently large and thus (by (5.3))

$$\| m_k \| \leqslant (r_0 + (1/n_k))(1 - \delta(t)).$$

Thus $\limsup_{k \to \infty} \| m_k \| \leqslant r_0(1 - \delta(t)) < r_0$. Since Proposition 10.1 implies $\lim_{k \to \infty} \| m_k - Tm_k \| = 0$, this contradicts the definition of r_0.

Proof of Theorem 10.4 Suppose $\{u_n\}$ in K satisfies w-$\lim_{n \to \infty} u_n = u$ while $\lim_{n \to \infty} \| u_n - Tu_n - w \| = 0$. Since these limits are preserved under the translation $x \mapsto x - w$, we may assume $w = 0$. Applying Proposition 10.2 to the set $K_n = \overline{\text{conv}}\{u_n, u_{n+1}, \ldots\}$, there exists $y_n \in K_n$ such that $y_n = Ty_n$. Since any weak subsequential limit of y_n lies in $\bigcap_{n=1}^{\infty} K_n = \{u\}$, it follows that w-$\lim_{n \to \infty} y_n = u$. Hence u is in the weak closure of the fixed point set, $F(T)$, of T. But since X is uniformly convex, X is both strictly convex and reflexive so $F(T)$ is closed and convex, hence weakly closed. Thus $u \in F(T)$, completing the proof.

Before stating the final theorem of this chapter we need some further definitions, as well as a fact about nonexpansive mappings defined in uniformly convex settings.

Definition 10.3 Let X be a (real) Banach space. The *normalized duality mapping* $J: X \to 2^{X^*}$ is defined by

$$J(z) = \{x^* \in X^* : \langle z, x^* \rangle = \| z \|^2 = \| x^* \|^2\} \qquad (z \in X).$$

Recall (Chapter 7) that a Banach space has Fréchet differentiable norm if for each $x \in X, x \neq 0$, the limit

$$D_x(y) = \lim_{t \to 0} t^{-1}(\| x + ty \| - \| x \|)$$

exists uniformly for bounded $y \in X$.

It is known (see Browder, 1976, p. 44) that for such X the duality mapping J is single-valued, and, moreover,

$$D_x(y) = \| x \|^{-1} \langle y, Jx \rangle.$$

Spaces with Fréchet differentiable norm include all the classical l^p, L^p spaces, $1 < p < \infty$.

Nonexpansive mappings, when restricted to convex subsets of uniformly convex spaces, satisfy a property which approximates linearity. This fact was first noticed by R.E. Bruck (1979) who described it as follows.

Let X be a Banach space and K a convex subset of X, and let Γ denote the set of strictly increasing convex continuous functions $\gamma \colon \mathbb{R}^+ \to \mathbb{R}^+$ with $\gamma(0) = 0$. A mapping $T \colon K \to X$ is said to be of *type* Γ if there exists $\gamma \in \Gamma$ such that for all $x, y \in K$ and $c \in [0, 1]$,

$$\gamma(\|cTx + (1-c)Ty - T(cx + (1-c)y)\|) \leqslant \|x - y\| - \|Tx - Ty\|.$$

Three facts about such mappings are obvious:

(a) Mappings of type Γ are nonexpansive.
(b) Affine nonexpansive mappings are of type Γ.
(c) Mappings of type Γ have convex fixed point sets.

Proposition 10.3 *If X is uniformly convex and if K is a convex subset of X, then there exists $\gamma \in \Gamma$ (depending on the diameter of K) such that any nonexpansive mapping $T \colon K \to X$ is of type Γ relative to γ.*

The following elementary fact about the modulus of convexity is used in the proof of the above proposition.

Lemma 10.1 *Let X be a Banach space with modulus of convexity δ. Then for all $u, v \in X$ with $\|u\| \leqslant 1$, $\|v\| \leqslant 1$, and all $c \in [0, 1]$,*

$$2 \min(c, 1-c)\delta(\|u - v\|) \leqslant 1 - \|cu + (1-c)v\|.$$

Proof Obviously we may suppose $c \leqslant \frac{1}{2}$. Since $\delta(\|u - v\|) \leqslant 1 - \frac{1}{2}\|u + v\|$, it suffices to show:

$$2c(1 - \frac{1}{2}\|u + v\|) \leqslant 1 - \|cu + (1-c)v\|,$$

i.e.,

$$2c + \|cu + (1-c)v\| \leqslant 1 + c\|u + v\|.$$

However this is a direct consequence of:

$$\begin{aligned}
2c + \|cu + (1-c)v\| &= 2c + \|c(u+v) + (1-2c)v\| \\
&\leqslant 2c + c\|u + v\| + (1-2c)\|v\| \\
&\leqslant 2c + c\|u + v\| + 1 - 2c \\
&= 1 + c\|u + v\|.
\end{aligned}$$

Proof of Proposition 10.3 Let δ denote the modulus of convexity of X. While we have seen that δ is continuous and strictly increasing (with $\delta(0)=0$), it is known that δ need not be convex (Example 5.8). However the function d defined as follows is in Γ.

$$d(t) = \begin{cases} \dfrac{1}{2}\displaystyle\int_0^t \delta(s)\,ds & \text{if } 0 \leqslant t \leqslant 2 \\[2mm] d(2)+\tfrac{1}{2}\delta(2)(t-2), & t > 2. \end{cases}$$

Note that for $t \in (0, 2]$,

$$0 < d(t) = \frac{1}{2}\int_0^t \delta(s)\,ds < (t/2)\delta(t) \leqslant \delta(t).$$

To see that d is convex note that for $t \in [0, 2]$, $d'(t) = \tfrac{1}{2}\delta(t)$; thus d' is increasing with $d'(2) = \tfrac{1}{2}\delta(2) = \tfrac{1}{2}$.

Now let $\|u\| \leqslant 1$, $\|v\| \leqslant 1$, and $c \in [0, 1]$. Then $c(1-c) \leqslant \min\{c, (1-c)\}$; thus by the lemma

$$2c(1-c)d(\|u-v\|) = \frac{2c(1-c)}{2}\int_0^{\|u-v\|} \delta(s)\,ds$$

$$\leqslant c(1-c)\delta(\|u-v\|)\|u-v\|$$
$$\leqslant 2c(1-c)\delta(\|u-v\|)$$
$$\leqslant 2\min\{c, 1-c\}\delta(\|u-v\|)$$
$$\leqslant 1-\|cu+(1-c)v\|. \tag{10.3}$$

Also for $ks < 2$ the function $s \mapsto d(ks)/s$ is increasing (since $(d(ks)/s)' = \{[\tfrac{1}{2}s\delta(ks)]\cdot k - d(ks)\}/s^2 \geqslant 0$).

Now let $D = \operatorname{diam} K$ and $x, y \in K$. Then for $c \in [0, 1]$,

$$c(1-c)\|x-y\| \leqslant D/4.$$

Set

$$u = [Ty - T(cx+(1-c)y)]/[c\|x-y\|];$$
$$v = [T(cx+(1-c)y) - Tx]/[(1-c)\|x-y\|].$$

Note that $\|u\| \leqslant 1$, $\|v\| \leqslant 1$; thus $\|u-v\| \leqslant 2$. From (10.3),

$$2c(1-c)\|x-y\|d(\|u-v\|) \leqslant \|x-y\| - \|Tx-Ty\|. \tag{10.4}$$

On the other hand,

$$d(\|u-v\|)=d\left(\left\|\frac{Ty-T(cx+(1-c)y)}{c\|x-y)\|}-\frac{T(cx+(1-c)y)-Tx}{(1-c)\|x-y\|}\right\|\right)$$

$$=d\left(\left\|\frac{cTx+(1-c)Ty-T(cx+(1-c)y)}{c(1-c)\|x-y\|}\right\|\right).$$

Taking $k=\|cTx+(1-c)Ty-T(cx+(1-c)y)\|$ and $s=[c(1-c)\|x-y\|]^{-1}$, $ks\leqslant 2$ and

$$2c(1-c)\|x-y\|d(\|u-v\|)=2d(ks)/s$$

$$\geqslant 2d(4k/D)(4/D)^{-1}$$

$$=(D/2)d(4k/D).$$

With (10.4), this yields

$$\tfrac{1}{2}Dd\left(\frac{4}{D}\|T(cx+(1-c)y)-cTx-(1-c)Ty\|\right)\leqslant\|x-y\|-\|Tx-Ty\|.$$

The fact that T is of type Γ may now be verified by taking $\gamma\in\Gamma$ as

$$\gamma(t)=\tfrac{1}{2}Dd\left(\frac{4t}{D}\right), \qquad t\geqslant 0.$$

We should remark that in (Bruck, 1981) Bruck uses Proposition 10.3 to obtain the following useful extension of Proposition 10.1. Let X be a Banach space, $K\subset X$, and $T: K\to K$. For $\mu>0$, set

$$F_\mu(T)=\{x\in K:\|x-Tx\|\leqslant\mu\}.$$

Theorem 10.5 (Bruck, 1981) *Suppose K is a bounded, closed and convex subset of a uniformly convex Banach space X. Then for each $\epsilon>0$ there exists $\sigma>0$, depending only on X,ϵ, and diam K, such that if $T: K\to K$ is nonexpansive, then*

$$\overline{\operatorname{conv}} F_\sigma(T)\subset F_\epsilon(T).$$

We now turn to the final theorem of this chapter. We saw in the previous chapter that if K is a bounded, closed and convex subset of X and if $T: K\to K$ is nonexpansive, then the mapping $S=\tfrac{1}{2}(I+T)$ is asymptotically regular. Thus if $x \in K$ and if $\{x_n\}$ is defined by

$$x_{n+1} = \tfrac{1}{2}(x_n + Tx_n) = Sx_n, \qquad n = 1, 2, \ldots,$$

then

$$\lim_{n \to \infty} \|x_n - Sx_n\| \equiv \lim_{n \to \infty} \|S^n x - S^{n+1} x\| = 0.$$

In view of Theorem 10.4, if X is also uniformly convex, then every weak subsequential limit of $\{x_n\}$ is always a fixed point of S (hence T). The following theorem, due to Reich (1979), shows that the additional assumption that X have Fréchet differentiable norm ensures that $\{x_n\}$ itself converges weakly.

Theorem 10.6 *Let K be a bounded, closed and convex subset of a uniformly convex Banach space X which also has Fréchet differentiable norm. Suppose $T: K \to K$ is nonexpansive, and set $S = \tfrac{1}{2}(I + T)$. Then for each $x \in K$, the sequence $\{S^n x\}$ converges weakly to a fixed point of T.*

Proof In view of the remarks preceding the theorem, it need only be shown that if w_1 and w_2 are weak subsequential limit points of $\{S^n x\}$ then $w_1 = w_2$.

For each $t \in [0, 1]$, let

$$a_n = a_n(t) = \|w_1 - w_2 + t(S^n x - w_1)\| - \|w_1 - w_2\|$$

and set

$$b_{n,m} = \|S^m(tS^n x + (1-t)w_1) - (tS^{n+m} x + (1-t)w_1)\|.$$

Then (since $S^m w_2 = w_2$)

$$\begin{aligned}
\|w_1 - w_2 + t(S^{n+m} x - w_1)\| &= \|w_1 - w_2 + S^m(tS^n x + (1-t)w_1) \\
&\quad - S^m(tS^n x + (1-t)w_1) + t(S^{n+m} x - w_1)\| \\
&\leqslant \|w_1 - S^m(tS^n x + (1-t)w_1) + t(S^{m+n} x - w_1)\| \\
&\quad + \|S^m(tS^n x + (1-t)w_1) - S^m w_2\| \\
&\leqslant \|S^m(tS^n x + (1-t)w_1) - (tS^{m+n} x + (1-t)w_1)\| \\
&\quad + \|w_1 - w_2 + t(S^n x - w_1)\|.
\end{aligned}$$

It follows that

$$a_{n+m} \leqslant b_{n,m} + a_n, \qquad m, n = 1, 2, \ldots . \tag{10.5}$$

Now let $M = \|x - w_1\|$. Since S^m is of type Γ on K, there exists $\gamma \in \Gamma$ such that

$$\gamma(\|S^m(tS^nx+(1-t)w_1)-(tS^mS^nx+(1-t)S^mw_1)\|)$$
$$\leqslant \|S^nx-w_1\|-\|S^{m+n}x-w_1\|,$$

i.e.,

$$\gamma(b_{n,m})\leqslant \|S^nx-w_1\|-\|S^{m+n}x-w_1\|.$$

Since $\{\|S^nx-w_1\|\}$ is monotone decreasing, this in turn implies that $\lim_{n\to\infty}\gamma(b_{n,m})=0$ uniformly for all m. This fact and (10.5) yield

$$\limsup_{n\to\infty} a_n\leqslant\liminf_{n\to\infty} a_n,$$

i.e.,

$$\lim_{n\to\infty} a_n(t)\equiv a(t)$$

exists for all $t\in[0,1)$.

Now (assuming $w_1\neq w_2$) let

$$d_n=\langle S^nx-w_1, J(w_1-w_2)/\|w_1-w_2\|\rangle.$$

Since the norm of X is Fréchet differentiable, if $\epsilon>0$ there exists $t=t(\epsilon)$ such that

$$|a_n(t)/t-d_n|<\epsilon.$$

It follows that

$$\limsup_{n\to\infty} d_n\leqslant a(t)/t+\epsilon \qquad \text{and} \qquad \liminf_{n\to\infty} d_n\geqslant a(t)/t-\epsilon;$$

hence $\lim_{n\to\infty} d_n$ exists and, since some subsequence of $\{S^nx\}$ converges weakly to w_1, it must be the case that $\lim_{n\to\infty} d_n=0$. Since the corresponding limit with w_1 and w_2 interchanged must also be 0, and since

$$\|w_1-w_2\|^2=\langle w_1-w_2, J(w_1-w_2)\rangle$$
$$=\langle w_1-S^nx, J(w_1-w_2)\rangle+\langle w_2-S^nx, J(w_2-w_1)\rangle,$$

it follows that $w_1=w_2$. This completes the proof.

Remark The above theorem is due to S. Reich (1979), who actually proved it with $\{S^nx\}$ replaced by the more general iterative process defined as follows:

$$x_1\in K;\ x_{n+1}=(1-c_n)x_n+c_nTx_n, \qquad n=1,2,\ldots,$$

where $0\leqslant c_n\leqslant 1$ and $\Sigma_{n=1}^{\infty} c_n(1-c_n)=+\infty$.

11

Properties of fixed point sets and minimal sets

We already know some facts about the structure of the fixed point sets of nonexpansive mappings. Lemma 3.4 states that if the space is strictly convex then such sets are always convex if the domain of the mapping is convex. In nonstrictly convex settings these sets need not be convex in the usual (linear space) sense. However, as we shall see, they are usually convex in a metric sense.

Recall that a metric space (M, ρ) is said to be *metrically convex* if for any two points $x, y \in M$ with $x \neq y$ there exists $z \in M$, $x \neq z \neq y$, such that

$$\rho(x, z) + \rho(z, y) = \rho(x, y).$$

As we have seen (Example 2.7), Menger's fundamental theorem on metric convexity asserts that in a complete and metrically convex metric space each two points are joined by a metric segment (a subset of the space which is isometric with a real line interval whose length is equal to the distance between the points). Examples of metric segments and metric 'lines' which are not algebraically linear arose in Example 3.6. Obviously convex subsets of Banach spaces are metrically convex, and in strictly convex Banach spaces these are the only metrically convex sets.

For a wide class of Banach spaces, the fixed point sets of nonexpansive mappings are metrically convex (hence, in particular always connected).

Definition 11.1 We shall say that a Banach space X has the *fixed point property* (*f.p.p.*) *for spheres* if each nonempty, closed and convex subset of the unit sphere has the f.p.p. for nonexpansive mappings.

Although the above definition may seem artificial, spaces having this property include all strictly convex spaces and, more generally, all spaces having Kadec–Klee norm. In the first instance the only convex sets lying on the unit sphere are singletons and in the second, all such sets are compact. The property isolated in the definition seems to have strong implications regarding the structure of fixed point sets of nonexpansive mappings.

As usual, suppose K is a closed and convex subset of a Banach space X, and suppose X has the f.p.p. for spheres. Let $T: K \to K$ be nonexpansive and let Fix T denote the fixed point set of T. We also suppose Fix $T \neq \emptyset$, and we assume K is 'locally' weakly compact – i.e., the intersection of K and any closed ball in X is weakly compact. Now let K_0 be any nonempty, closed and convex subset of K which is invariant under T. Then Fix $T \cap K_0 \neq \emptyset$. To see this, let $x \in$ Fix T and consider

$$K_1 = \mathrm{Proj}_{K_0}(x) = \{y \in K_0: \|x - y\| = \mathrm{dist}(x, K_0)\}$$
$$= B(x; \mathrm{dist}(x, K_0)) \cap K_0.$$

Then K_1 is a closed and convex subset of a sphere which, by weak compactness, is nonempty. Also, for any $y \in K_1$, $\|Ty - x\| = \|Ty - Tx\| \leqslant \|y - x\|$; thus $T: K_1 \to K_1$ and by the f.p.p. for spheres,

$$\text{Fix } T \cap K_0 \supset \text{Fix } T \cap K_1 \neq \emptyset.$$

Therefore we have:

Proposition 11.1 *Let X be a Banach space which has the f.p.p. for spheres, let K be a bounded, closed, convex subset of X, and let $T: K \to K$ be nonexpansive. Then:*

> *Either* Fix $T = \emptyset$, *or T has fixed points in any nonempty, closed, convex, T-invariant subset of K.* (CFP)

Mappings satisfying conclusion (CFP) are said to have the *conditional f.p.p.* This property was introduced and studied by Bruck in 1973.

Theorem 11.1 *Suppose K is a nonempty, closed, convex subset of a Banach space having the f.p.p. for spheres and suppose $T: K \to K$ is nonexpansive. Then* Fix T *is metrically convex.*

Proof Observe that if $x, y \in$ Fix T with $\|x - y\| = d = d_1 + d_2 > 0$, then the set

$$K_1 = B(x; d_1) \cap B(y; d_2)$$

is a closed and convex subset of a sphere. It follows from the triangle inequality and nonexpansiveness of T that $T: K_1 \to K_1$. Thus

$$\text{Fix } T \cap K_1 \neq \emptyset.$$

A second interesting property of the fixed point sets of nonexpansive mappings involves the notion of retraction. A subset H of K is said to be a *retract* of K if there exists a continuous mapping $R: K \to H$ with Fix $R = H$. Any such mapping R is a *retraction* of K onto H. If R is nonexpansive, then H is said to be a *nonexpansive retract* of K. (Recall that in Example 3.6 we presented several nonexpansive retracts in the space $\mathbb{R}^2$ with l_2^∞ norm.)

The following theorem is due to Bruck (1973).

Theorem 11.2 *Let X be a Banach space and K be a nonempty, locally weakly compact, convex subset of X and suppose $T: K \to K$ is a nonexpansive mapping which satisfies (CFP). Then Fix T is a nonexpansive retract of K.*

We derive Theorem 11.2 from the following three lemmas.

Lemma 11.1 *Suppose K is convex and locally weakly compact, and let F be a nonempty subset of K. Then the set*

$$N(F) = \{f \in K^K : f \text{ is nonexpansive and Fix } f \supset F\}$$

is compact in the topology of weak pointwise convergence.

Proof Fix $x_0 \in F$ and for $x \in K$ set

$$K_x = \{y \in K : \|y - x_0\| \leqslant \|x - x_0\|\}.$$

Note that if $f \in N(F)$ then $f(x) \in K_x$ for each $x \in K$, i.e., f is an element of the Cartesian product $P = \Pi_{x \in K} K_x$. Since each set K_x is a bounded, closed and convex subset of K, each such set is compact in the weak topology, so by Tychonoff's theorem P is compact in the product topology, i.e., the topology of weak pointwise convergence. Since $N(F) \subset P$, it need only be shown that $N(F)$ is closed in the topology of weak pointwise convergence. Suppose $\{f_\alpha, \alpha \in A\}$ is a net in $N(F)$ for which $w\text{-}\lim_\alpha f_\alpha x = fx$, $x \in K$. Then clearly $fx = x$ for all $x \in F$ and moreover if $u, v \in K$,

$$\|fu - fv\| \leqslant \liminf_\alpha \|f_\alpha u - f_\alpha v\| \leqslant \|u - v\|.$$

Thus $f \in N(f)$, completing the proof.

Lemma 11.2 *Let K and F be as in Lemma 11.1. Then for some $r \in N(F)$,*

$$\|(fr)x - (fr)y\| = \|rx - ry\|$$

for all $x, y \in K$, $f \in N(F)$.

Proof Introduce a partial order (preorder) $\leqslant$ in $N(F)$ by setting for $f, g \in N(F)$, $f \leqslant g$ if $\|fx - fy\| \leqslant \|gx - gy\|$ for all $x, y \in K$. To see that $N(F)$ has a minimal element, let $\{f_\lambda, \lambda \in \Lambda\}$ be linearly ordered in $(N(F), \leqslant)$. Then the sets

$$I_\lambda = \{g \in N(F) : g \leqslant f_\lambda\}$$

are linearly ordered by set inclusion. Moreover, as in Lemma 11.1, each I_λ is closed, hence compact, in the topology of weak pointwise convergence.

Thus $\bigcap_{\lambda \in \Lambda} I_\lambda \neq \emptyset$, i.e., there exists $g \in \bigcap_{\lambda \in \Lambda} I_\lambda$ such that $g \in N(F)$ and $g \leqslant g_\lambda$ for all $\lambda \in \Lambda$.

We may now invoke Zorn's Lemma and conclude that $(N(F), \leqslant)$ has a minimal element r. Since $N(F)$ is closed under functional composition and the mappings are nonexpansive, the conclusion follows.

Lemma 11.3 *Let K and F be as in Lemma 11.1, and suppose that for each $z \in K$ there exists $h \in N(F)$ such that $hz \in F$. Then F is a nonexpansive retract of K.*

Proof Let r be the nonexpansive mapping of Lemma 11.2. Since $r \in N(F)$ it suffices to show that under the stated hypotheses r maps K into F. Let $p \in K$ and set $z = rp$. By assumption there exists $h \in N(F)$ such that $y = (hr)p \in F$. By Lemma 11.2,

$$\|(hr)p - (hr)y\| = \|rp - ry\|.$$

But since $y \in F$, $y = (hr)y = (hr)p$ and the above equality implies $rp = ry$. Finally, $ry = y$ since $y \in F$, thus $rp = ry = y \in F$.

We now return to the theorem.

Proof of Theorem 11.2 We may assume Fix $T \neq \emptyset$. Let $z \in K$ and set

$$C = \{fz : f \in N(\text{Fix } T)\}.$$

Then C is the image of $N(\text{Fix } T)$ in the product space P of Lemma 11.1 under the zth coordinate projection. Since $N(\text{Fix } T)$ is compact in P (Lemma 11.1), it follows that C must be weakly compact, hence bounded. Also, since Fix $T \neq \emptyset$, clearly $C \neq \emptyset$. Moreover, C is obviously convex and, since $Tf \in N(\text{Fix } T)$ whenever $f \in N(\text{Fix } T)$, $T : C \to C$. It follows from (CFP) that T has a fixed point in C, i.e., there exists $h \in N(\text{Fix } T)$ with $hz \in \text{Fix } T$. Since $z \in K$ was arbitrary the hypothesis of Lemma 11.3 is satisfied and Fix T is a nonexpansive retract of K.

Theorem 11.2 is not true if the assumption (CFP) is dropped. In fact, we have already seen that Fix T may even be disconnected (Example 3.7). However the analogue of Theorem 11.2 does hold for hyperconvex spaces (Baillon, 1988).

With K and T as above, if Fix $T \neq \emptyset$ then Fix T always intersects the asymptotic centers of its sequences. This is because any sequence $\{x_n\}$ in Fix T is an approximate fixed point sequence; hence $A(K, \{x_n\})$ is T-invariant. In particular, if X is uniformly convex then $A(K, \{x_n\})$ is always a singleton. Thus in such spaces, Fix T is 'closed' with respect to its asymptotic centers, i.e., if $\{x_n\} \subset \text{Fix } T$ then $A(K, \{x_n\}) \in \text{Fix } T$. This fact can be used to show that some (even convex) sets cannot be fixed point sets for any nonexpansive mapping and hence cannot be nonexpansive retracts.

Example 11.1 Consider the space l_3^p (the space $\mathbb{R}^3$ furnished with the norm $\|(x_1, x_2, x_3)\|_p = (|x_1|^p + |x_2|^p + |x_3|^p)^{1/p})$ for $1 < p < \infty$, and let E denote the plane $\{x = (x_1, x_2, x_3) \in \mathbb{R}^3 : x_1 + x_2 + x_3 = 1\}$. Then if $p \neq 2$, E cannot be a nonexpansive retract of l_3^p because, in particular, there are no nonexpansive $T: l_3^p \to l_3^p$ for which $\text{Fix } T = E$. To see this, consider the basis vectors $e^1 = (1, 0, 0)$, $e^2 = (0, 1, 0)$, $e^3 = (0, 0, 1)$, and let $\{x_n\}$ be the periodic sequence $\{e^1, e^2, e^3, e^1, e^2, e^3, \ldots\}$. Then the asymptotic center of $\{x_n\}$ in l_3^p coincides with the Chebyshev center of the set $\{e^1, e^2, e^3\}$, and this is easily seen to be the point (c, c, c) with $c = (1 + 2^{(1/p-1)})^{-1}$. This point belongs to E only when $p = 2$. Obviously in this case E is a nonexpansive retract of l_3^2 via the orthogonal projection of l_3^2 onto E.

Our next observation regarding the structure of the fixed point sets is related to a subclass of the class of nonexpansive mappings. Let K be convex, let $T: K \to K$ be any mapping, and for $x, y \in K$ consider the function $\Phi_{x,y}$ defined by

$$\Phi_{x,y}(t) = \|(1-t)(x-y) + t(Tx - Ty)\|, \qquad t \in [0, 1].$$

Obviously, $\Phi_{x,y}$ is a convex function of t.

Definition 11.2 With K as above, a mapping $T: K \to K$ is said to be *firmly nonexpansive* if for any $x, y \in K$ the function $\Phi_{x,y}$ is nonincreasing on $[0, 1]$.

Although any firmly nonexpansive mapping must be nonexpansive ($\Phi_{x,y}(0) \geq \Phi_{x,y}(1)$), the converse need not be true. (Consider the mapping $Tx = -x$ in any space X.) Also, since $\Phi_{x,y}$ is convex it is easy to check that a mapping $T: K \to K$ is firmly nonexpansive if and only if:

$$\|Tx - Ty\| \leq \|(1-t)(x-y) + t(Tx - Ty)\|, \qquad x, y \in K, t \in [0, 1]. \quad (11.1)$$

In view of the above one might expect firmly nonexpansive mappings to exhibit better behavior than nonexpansive mappings in general. However, from the point of view of fixed point theory, the restriction is mild.

Theorem 11.3 *Let K be a nonempty, closed, convex subset of a Banach space X and let $T: K \to K$ be nonexpansive. Then there exists a family $F_\alpha: K \to K$, $\alpha \in [0, 1)$, of firmly nonexpansive mappings with the following properties:*

(a) *Any closed, convex, T-invariant subset K_0 of K is F_α-invariant for all* $\alpha \in [0, 1)$.
(b) Fix $F_\alpha = $ Fix T *for all* $\alpha \in [0, 1)$.
(c) *If T has bounded orbits then for any* $x \in K$, $\lim_{\alpha \to 1} \| F_\alpha x - T F_\alpha x \| = 0$.

Proof Select $x \in K, \alpha \in [0, 1)$, and consider the mapping $T_x : K \to K$ defined by

$$T_x z = (1 - \alpha)x + \alpha T z, \qquad z \in K. \tag{11.2}$$

The mapping T_x has Lipschitz constant $\alpha < 1$, so by Banach's Contraction Principle there exists exactly one point $F_\alpha x \in K$ such that $T_x F_\alpha x = F_\alpha x$. This in turn defines a mapping $F_\alpha : K \to K$ and F_α is nonexpansive. To see this, let $u, v \in K$. Then in view of (11.2):

$$F_\alpha u = (1 - \alpha)u + \alpha T F_\alpha u,$$
$$F_\alpha v = (1 - \alpha)v + \alpha T F_\alpha v.$$

Thus

$$\| F_\alpha u - F_\alpha v \| \leqslant (1 - \alpha)\| u - v \| + \alpha \| T F_\alpha u - T F_\alpha v \|$$
$$\leqslant (1 - \alpha)\| u - v \| + \alpha \| F_\alpha u - F_\alpha v \|$$

implying

$$\| F_\alpha u - F_\alpha v \| \leqslant \| u - v \|.$$

Property (a) is obvious. Also (b) is immediate since $F_\alpha x = x$ implies $x = (1 - \alpha)x + \alpha T x$. To check (c) suppose T has bounded orbits. Then a routine argument shows that K contains a *bounded*, convex, T-invariant subset K_0, and moreover K_0 may be chosen in such a way that it contains any prescribed $x \in K$. It follows that $\{ F_\alpha x \} \subset K_0$ and thus $\{ F_\alpha x \}$ is bounded for $\alpha \in [0, 1)$. Property (c) now follows from the equality:

$$\| T F_\alpha x - F_\alpha x \| = \frac{1 - \alpha}{\alpha} \| F_\alpha x - x \|$$

which is an easy consequence of the fact:

$$F_\alpha x = (1 - \alpha)x + \alpha T F_\alpha x. \tag{11.3}$$

It remains to be shown that each F_α is firmly nonexpansive. Fix $\alpha \in [0, 1)$ and take any $t \in (0, 1)$. Set $p = (1 - t)x + t F_\alpha x, q = (1 - t)y + t F_\alpha y$.

Using (11.3),

$$F_\alpha x = (1-\alpha)\left(\frac{-t}{1-t}F_\alpha x + \frac{1}{1-t}p\right) + \alpha T F_\alpha x,$$

$$F_\alpha x = \left(1 - \alpha\frac{1-t}{1-\alpha t}\right)p + \alpha\frac{1-t}{1-\alpha t}T F_\alpha x.$$

This shows that $F_\alpha x = F_\beta p$ for $\beta = \alpha(1-t)/(1-\alpha t)$. Similarly, $F_\alpha y = F_\beta q$, so

$$\|F_\alpha x - F_\alpha y\| = \|F_\beta p - F_\beta q\|$$

$$\leqslant \|p - q\|$$

$$= \|(1-t)(x-y) + t(F_\alpha x - F_\alpha y)\|.$$

Thus (11.1) holds, proving T is firmly nonexpansive.

The above theorem shows that one may study the structure of fixed point sets of nonexpansive mappings by restricting one's attention to only those sets which are also fixed point sets of firmly nonexpansive mappings.

The passing from T to F_α is a regularizing technique which produces an 'approximating curve' $\{F_\alpha x\}_{\alpha \in (0,1)}$ which, in view of (c), is analogous to an approximating fixed point sequence. However it should be noted that, in general, $\lim_{\alpha \to 1} F_\alpha x$ does not exist. This can be seen by analyzing Example 3.7. The strongest result known concerning the above limit is one due to S. Reich (1980) which states that $F_\alpha x$ converges to a fixed point of T provided the underlying space is uniformly smooth. We shall examine the simpler Hilbert space case in the next chapter.

Finally we observe that F_α is a convex combination of the identity mapping and the nonexpansive mapping TF_α. Therefore, by Theorem 9.4, F_α is always asymptotically regular for $\alpha \in (0, 1)$ and moreover, in view of (11.2) and (11.3), F_α has the following representation:

$$F_\alpha = (I - \alpha T)^{-1}(1-\alpha)I. \tag{11.4}$$

We now turn to an analysis of the structure of minimal invariant convex sets consisting of more than one point. While any nonempty, weakly compact, convex, T-invariant set always contains a nonempty minimal closed convex T-invariant subset, it was not until the 1981 example of D. Alspach that it became known that such sets could have positive diameter. (In this connection, also see Schechtman, 1982.)

Example 11.2 (Alspach's) Let $X = L^1[0, 1]$. All equations and inequalities which follow should be thought of in the L^1 sense, i.e., as holding almost everywhere. It is known that in L^1 any set of the form $\{f : g \leqslant f \leqslant h\}$ with g and h fixed (which is called an order interval) is weakly compact. Consider the set

$$K = \left\{ f \in L^1 : \int_0^1 f = 1, 0 \leqslant f \leqslant 2 \right\}.$$

This set is a weakly closed subset of an order interval and thus is itself weakly compact. Clearly K is also convex and diam $K = 2$. Define the isometry $T : K \to K$ by

$$(Tf)(t) = \begin{cases} \min\{2f(2t), 2\} & \text{if } 0 \leqslant t \leqslant \tfrac{1}{2} \\ \max\{2f(2t-1) - 2, 0\} & \text{if } \tfrac{1}{2} < t \leqslant 1. \end{cases} \tag{11.5}$$

Suppose $Tg = g$ for some $g \in L^1$. Then

$$\{t : g(t) = 2\} = \{t : (Tg)(t) = 2\},$$

and in view of (11.5),

$$\{t : g(t) = 2\} = \{t/2 : g(t) = 2\} \cup \left\{ \frac{1+t}{2} : g(t) = 2 \right\} \cup \{t/2 : 1 \leqslant g(t) < 2\}.$$

Since the above sets are disjoint and since the measure of

$$\{t/2 : g(t) = 2\} \cup \left\{ \frac{1+t}{2} : g(t) = 2 \right\}$$

is equal to the measure of $\{t : g(t) = 2\}$, the measure of the set $\{t/2 : 1 \leqslant g(t) < 2\}$ is zero. Continuing this argument it follows that the set

$$\{t : 0 < g(t) < 2\} = \bigcup_{n=0}^{\infty} \{t : 2^{-n} \leqslant g(t) < 2^{-n+1}\}$$

has measure zero as well, and thus there exists a set $A \subset [0, 1]$ whose characteristic function χ satisfies $g(t) = 2\chi_A(t)$. In view of the construction of T, A must have measure $\tfrac{1}{2}$. Also, since $Tg = g$, A must be 'similar' to its intersection with each of the intervals $[0, \tfrac{1}{2}]$ and $[\tfrac{1}{2}, 1]$, and the same is true of any interval of the form $[k/2^n, (k+1)/2^n]$. It follows that the intersection of A with any of the above intervals has measure exactly half the measure of the interval and this in turn implies the same is true for any interval, contradicting the Banach Density Theorem (which states that any measurable set has density either zero or one almost everywhere).

The set K in the above example is not minimal T-invariant. In fact, as we have seen earlier, weakly compact convex T-invariant sets must be diametral. However diam $K = 2$ while the Chebyshev radius $r(K) = 1$ and the function $f \equiv 1$ is the unique point in the Chebyshev center of K. On the other hand, K must contain a minimal nonempty, T-invariant, weakly compact, convex set. It is interesting to note that so far no explicit description of such a set has been found.

Now let K be an *arbitrary* minimal, nonempty, convex, T-invariant set for some nonexpansive mapping T. We have already seen that K must have the following two properties:

Property 11.1 $K = \overline{\operatorname{conv}}\{T(K)\}$.

Property 11.2 *K is diametral.*

We now list several additional properties any such set K must have.

Property 11.3 *For fixed* $x \in K$, $r(z, \{T^n x\}) \equiv \lim_{n \to \infty} \sup \|z - T^n x\|$ *is constant for* $z \in K$. *Thus* $r(z, \{T^n x\}) = r(K, \{T^n x\})$ *and* $A(K, \{T^n x\}) = K$.

Indeed, if $r(z, \{T^n x\})$ were not constant there would be a nontrivial subset of K (a level set) invariant under T.

A stronger version of Property 11.3 holds when K is weakly compact.

Property 11.4 *If K is weakly compact then for any $x, z \in K$.*

$$r(z, \{T^n x\}) = \operatorname{diam} K.$$

To verify Property 11.4, observe that if there exists $x, z \in K$ for which $r(z, \{T^n x\}) = r < \operatorname{diam} K$ then the family of all balls centered in K with radius $(\frac{1}{2})(r + d)$ would have the finite intersection property; hence these balls would have nonempty intersection. Any point in this intersection would be a nondiametral point of K, violating Property 11.2

Now let $\{y_n\}$ be an approximate fixed point sequence for T in K. Then $\lim_{n \to \infty} \|y_n - T y_n\| = 0$ and any subsequence of $\{y_n\}$ is again an approximate fixed point sequence. This fact, combined with observations similar to those in the preceding paragraphs, yields:

Property 11.5 *For any* $x \in K, r(x, \{y_n\}) = \lim_{n \to \infty} \|x - y_n\| \equiv r(K, \{y_n\})$.

Property 11.6 *If K is weakly compact then for any $x \in K$,*

$$r(x, \{y_n\}) = \lim_{n \to \infty} \|x - y_n\| = \operatorname{diam} K.$$

This last property commonly known as Karlovitz's Lemma, was discovered independently by Karlovitz (1976) and Goebel (1975b). A. Khamsi (1987) has recently proved that an analogous result holds for minimal T-invariant, weak* compact sets in certain dual Banach spaces.

We remark that no example is known where $r(K, \{y_n\}) < \operatorname{diam} K$ in Property 11.5.

The next two properties follow easily.

Property 11.7 *If K is weakly compact then K cannot be covered with a finite family of sets each having diameter less than* $\operatorname{diam} K$.

Property 11.8 *If K is weakly compact then K cannot be covered with a finite number of balls each having radius smaller than* $\operatorname{diam} K$.

In other words, for K minimal nonempty and convex,

$$\alpha(K) = \chi(K) = \operatorname{diam} K$$

where α and χ are, respectively, the Kuratowski and Hausdorff measures of noncompactness.

The above properties can sometimes be used to eliminate the possibility that certain weakly compact sets are minimal.

Example 11.3 Let $X = c_0$ and set $K = \overline{\operatorname{conv}}\{e^i\}$ where $\{e^i\}$ is the standard unit vector basis. Thus

$$K = \left\{ x = \{x_i\} : x_i \geqslant 0, \sum_{i=0}^{\infty} x_i \leqslant 1 \right\}.$$

Then K is weakly compact, and it is not difficult to see that K is diametral with $\operatorname{diam} K = 1$ and that $\alpha(K) = \chi(K) = 1$. Thus K satisfies Properties 11.2, 11.7, 11.8. However K cannot be a minimal T-invariant set for any nonexpansive T. To see this, suppose $T: K \to K$ is nonexpansive and construct the approximating set $\{F_\alpha 0\}_{\alpha \in (0,1)}$ as in (11.3). Then

$$\lim_{\alpha \to 1} \|F_\alpha 0 - T F_\alpha 0\| = 0$$

and by Property 11.6, $\lim_{\alpha \to 1} \|F_\alpha 0 - x\| = 1$ for all $x \in K$. In particular, $\lim_{\alpha \to 1} \|F_\alpha 0 - 0\| = 1$. Select $r \in (\frac{1}{2}, 1)$. Since $F_\alpha 0$ is continuous with respect to α there exists α_0 such that $\|F_\alpha 0\| \geqslant r$ if $\alpha > \alpha_0$. Therefore $F_\alpha 0 \in K \backslash B(0; r)$ for $\alpha > \alpha_0$. But the set $K \backslash B(0; r)$ consists of infinitely many piecewise connected disjoint components of the form

$$K \cap B(e^i; 1 - r)$$

each having diameter $1 - r$, and the curve $\{F_\alpha 0\}_{\alpha > \alpha_0}$ must remain in one of these components, violating Property 11.6.

12

Special properties of Hilbert space

Since the geometry of Hilbert space is euclidean, and thus the 'nicest' of that of all Banach spaces, it is natural to expect that the behavior of the nonexpansive mappings defined in this space would be more predictable than in general. The results of this chapter support this.

Let H denote a (real) Hilbert space, and let K be a nonempty, closed and convex subset of H. As usual, we denote the inner product of $x, y \in H$ by $\langle x, y \rangle$.

As we have already seen, H is uniformly convex with modulus of convexity δ_H given by

$$\delta_H(\epsilon) = 1 - (1 - (\epsilon/2)^2)^{1/2}.$$

Thus for any $x \in H$ the nearest point projection $\mathrm{Proj}_K(x)$ is well defined as the unique point of K which is nearest x; that is, $y = \mathrm{Proj}_K(x)$ if and only if $y \in K$ and $\|x - y\| = \inf\{\|x - z\| : z \in K\}$. We list some properties of Proj_K below.

Lemma 12.1 $y = \mathrm{Proj}_K(x)$ *if and only if for each* $z \in K$,

$$\langle z - y, y - x \rangle \geq 0. \tag{12.1}$$

Proof Let $y = \mathrm{Proj}_K(x)$ and consider the function $\psi : [0, 1] \to \mathbb{R}^+$ defined for fixed $z \in K$ by

$$\psi(t) = \|x - (1 - t) y - tz\|^2.$$

Then ψ is convex with $\psi(0) \leq \psi(t)$ for $t \in [0, 1]$; thus $\psi'(0) \geq 0$. But

$$\psi(t) = \|x - y\|^2 + 2t\langle y - x, z - y \rangle + t^2 \|y - z\|^2,$$

so

$$\psi'(0) = 2\langle y - x, z - y \rangle \geq 0. \tag{12.2}$$

On the other hand, reversing the above argument, (12.2) yields $\psi(0) \leq \psi(1)$; hence $\|x - y\| \leq \|x - z\|$.

Thus it is always the case that

$$\langle \mathrm{Proj}_K x - x, z - \mathrm{Proj}_K x \rangle \geq 0 \qquad \text{for all } z \in K. \tag{12.3}$$

126

We now assume K is fixed and simplify the notation by letting $P = \text{Proj}_K$. By (12.3), for any two points $x, y \in H$,

$$\langle Px - x, Py - Px \rangle \geqslant 0;$$
$$\langle Py - y, Px - Py \rangle \geqslant 0.$$

This implies

$$\langle (x-y) - (Px - Py), Px - Py \rangle \geqslant 0, \tag{12.4}$$

and thus

$$\|Px - Py\|^2 \leqslant \langle Px - Py, x - y \rangle \leqslant \|Px - Py\| \|x - y\|. \tag{12.5}$$

Therefore $\|Px - Py\| \leqslant \|x - y\|$, i.e., P is nonexpansive. However, (12.4) and (12.5) show more.

Definition 12.1 A mapping $T: D \to H$ is called *monotone* if

$$\langle Tx - Ty, x - y \rangle \geqslant 0, \qquad x, y \in D,$$

and *strictly monotone* if for some $c > 0$,

$$\langle Tx - Ty, x - y \rangle \geqslant c \|x - y\|^2.$$

In view of the above, both P and $(I - P)$ are monotone.

We now characterize all *firmly nonexpansive* mappings in H. Let $T: K \to H$ and consider $\varphi: [0, 1] \to \mathbb{R}^+$ defined by

$$\varphi(t) = \|(1-t)(x-y) + t(Tx - Ty)\|^2.$$

Since φ is convex, a necessary and sufficient condition for T to be firmly nonexpansive is that $\varphi'(1) \leqslant 0$. But

$$\varphi(t) = (1-t)^2 \|x-y\|^2 + 2t(1-t)\langle Tx - Ty, x - y \rangle + t^2 \|Tx - Ty\|^2,$$

so

$$\varphi'(t) = -2(1-t)\|x-y\|^2 + 2(1-2t)\langle Tx - Ty, x - y \rangle + 2t\|Tx - Ty\|^2,$$

and $\varphi'(1) \leqslant 0$ is equivalent to

$$\langle Tx - Ty, x - y \rangle \geqslant \|Tx - Ty\|^2. \tag{12.6}$$

In view of (12.5), the projections P are all firmly nonexpansive.

Now suppose T satisfies (12.6), and consider the mapping $S = 2T - I$. Then for $x, y \in K$,

$$\begin{aligned}
\|Sx - Sy\|^2 &= \|2Tx - x - 2Ty + y\|^2 \\
&= \|2(Tx - Ty) - (x - y)\|^2 \\
&= 4\|Tx - Ty\|^2 - 4\langle Tx - Ty, x - y \rangle + \|x - y\|^2 \\
&\leqslant \|x - y\|^2.
\end{aligned}$$

This shows that S is nonexpansive, so any firmly nonexpansive mapping T is of the form

$$T = \tfrac{1}{2}\,(I + S)$$

with S nonexpansive. Reversing this argument shows that mappings of the above form are always firmly nonexpansive.

We summarize these facts as follows.

Theorem 12.1 *For a mapping* $T: K \to H$ *the following are equivalent.*

(a) *T is firmly nonexpansive.*
(b) *$2T - I$ is nonexpansive.*
(c) *$T = \tfrac{1}{2}\,(I + S)$ with S nonexpansive.*
(d) $\langle Tx - Ty, x - y \rangle \geqslant \| Tx - Ty \|^2, x, y \in K.$

Theorem 12.2 *Any nonempty, closed and convex subset K of a Hilbert space H is a nonexpansive retract of H via the nonexpansive retraction $P = \mathrm{Proj}_K$. Moreover, P is firmly nonexpansive and monotone, $I - P$ is nonexpansive and monotone, and the reflection of H in K defined by $S = 2P - I$ is nonexpansive.*

In spaces other than Hilbert space the nearest point projections, even if well defined, are not in general nonexpansive.

Example 12.1 For any Banach space X the radial projection R of X onto the unit ball $B(0; 1)$ of X is given for $x \in X$:

$$Rx = \begin{cases} x & \text{if } \|x\| \leqslant 1 \\ x/\|x\| & \text{if } \|x\| > 1. \end{cases}$$

The mapping R is always a nearest point projection, and simple calculations show that for $x, y \in X$ with $\max\{\|x\|, \|y\|\} \geqslant 1$,

$$\| Rx - Ry \| \leqslant \frac{2}{\max\{\|x\|, \|y\|\}} \| x - y \| \leqslant 2 \| x - y \|.$$

Obviously $R = I$ on $B(0; 1)$. Thus $k(R) \leqslant 2$ for any space X. Since for Hilbert space $k(R) = 1$, the question of the value of $k(R)$ for other concrete spaces arises, and this is a technical problem about which little is known. It is easy to check, however, that for the spaces $l^1, l^\infty, L^1\,[0, 1], c_0$, and $C[0, 1], k(R) = 2$. Also, Thele (1974) has proved that $k(R) < 2$ if and only if X is uniformly nonsquare.

It might seem natural to think that the radial projection R is the only nearest point projection onto the unit ball of a Banach space, but this is not the case. If $X = C[0, 1]$ the mapping T given by taking for $x \in X$,

$$(Tx)(t) = \max\{-1, \min\{1, x(t)\}\}, \qquad t \in [0, 1],$$

is a nearest point projection of X onto $B(0; 1)$ but $T \neq R$. (Moreover, T is nonexpansive!)

Another special Hilbert space property is connected with approximating curves. Let K be a closed convex set in a Hilbert space H with $T: K \to K$ nonexpansive and $\emptyset \neq \text{Fix } T = K_0$. In particular, K_0 is closed and convex. For any $x \in K$ the approximating curve (Chapter 11) beginning at x is given by

$$F_\alpha x = (1 - \alpha)x + \alpha T F_\alpha x, \qquad \alpha \in [0, 1). \tag{12.7}$$

Let $d = \text{dist}(x, K_0)$ and let $y = \text{Proj}_{K_0} x$. Since F_α is firmly nonexpansive, $G_\alpha = 2F_\alpha - I$ is nonexpansive and, since $y = F_\alpha y = G_\alpha y = Ty$, we have

$$\|x - F_\alpha x\| = \tfrac{1}{2}\|x - G_\alpha x\|$$

$$\leqslant \tfrac{1}{2}(\|x - y\| + \|y - G_\alpha x\|)$$

$$= \tfrac{1}{2}(d + \|G_\alpha y - G_\alpha x\|)$$

$$\leqslant \tfrac{1}{2}(d + d)$$

$$= d.$$

We have already seen that $\lim_{\alpha \to 1^-} \|F_\alpha x - T F_\alpha x\| = 0$. Thus for any sequence $\{\alpha_n\}$ with $\alpha_n \uparrow 1$, the sequence $\{z_n\} = \{F_{\alpha_n} x\}$ is an approximate fixed point sequence for T. Since the only fixed point of T in the ball $B(x; d)$ is y, any weakly convergent subsequence of $\{z_n\}$ has y as its weak limit (and asymptotic center). Therefore $\lim_{n \to \infty} \|z_n - x\| = d$ and consequently

$$\lim_{n \to \infty} z_n = y$$

in the strong sense. Therefore we have the following.

Theorem 12.3 *If K is a closed convex subset of a Hilbert space with $T: K \to K$ a nonexpansive mapping for which* Fix $T \neq \emptyset$, *and if F_α is given by* (12.7), *then*

$$\text{Proj}_{\text{Fix } T} x = \lim_{\alpha \to 1^-} F_\alpha x, \qquad x \in K.$$

A second use of approximating curves is given by the following example, which is of interest when K is unbounded.

Example 12.2 With K and F_α as above it is easy to check that

$$\langle Tx - x, F_\alpha x - x \rangle \geqslant 0, \qquad x \in K. \tag{12.8}$$

We say that K is bounded at x in the direction $u, u \neq 0$, if the set $\{y \in K : \langle y - x, u \rangle \geqslant 0\}$ is bounded. It is easy to see that if a set K is bounded in the direction $u \neq 0$ for a given $x \in K$ for which $x + \lambda u \in K$ for small $\lambda > 0$, then it is also bounded in the direction u for each $x \in K$. Moreover, if K is bounded in the direction $Tx - x$ for at least one $x \in K$, then in view of (12.8) K must contain a bounded approximating curve and thus a fixed point of T.

For example, if K is the first quadrant of the euclidean plane and if for some $x \in K$, $Tx - x$ belongs to the open third quadrant (i.e., $T(x_1, x_2) = (y_1, y_2)$ with $y_1 < x_1, y_2 < x_2$), then T has a fixed point in K. (This observation illustrates another fact. Fixed point theorems for nonexpansive mappings do not always reduce to fixed point theorems for continuous mappings in finite dimensional settings.)

The unbounded, closed, convex sets in Banach spaces fall into two distinct categories, those which are linearly unbounded in the sense that they have an unbounded intersection with at least one line, and those which, in contrast, are linearly bounded. If K is linearly unbounded then it contains at least one 'half-line': $\{x(t) = x_0 + tu : x_0 \in K, u \neq 0, t \geqslant 0\}$. In this case K is invariant under the isometry (translation) T given by $Tx = x + u$ with $\|x - Tx\| \equiv \|u\| > 0$.

Any convex set K in a reflexive Banach space which is unbounded yet linearly bounded has the so-called 'almost fixed point property': $\inf\{\|x - T\| : x \in K\} = 0$ for each nonexpansive mapping $T : K \to K$. Special cases of this result were discovered by Goebel and Kuczumow (1978b) and Ray (1980), and it was established in full generality by Reich (1983). Our next examples illustrate this fact in the special Hilbert space setting.

Example 12.3 Let $H = l^2$ and consider the 'block' set

$$K = \{x = \{x_i\} : |x_i| \leqslant M_i\},$$

where $\{M_i\}$ is a given sequence of nonnegative real numbers with $\Sigma_{i=1}^{\infty} M_i^2 = +\infty$. Thus K is a closed, convex, unbounded set which is linearly bounded. For any $u \in l^2$ with $\|u\| = 1$, consider the cone

$$C = \{x \in l^2 : \langle x/\|x\|, u \rangle \geqslant a\}$$

with $a \in (0, 1]$ given. Then $C \cap K$ is bounded. Indeed, suppose there exists a sequence $\{y^n\}$ in $C \cap K$ with $\lim_{n \to \infty} \|y^n\| = +\infty$. Let H_0 be a subspace of l^2 spanned by a finite number of elements of the natural basis $\{e^i\} = \{\delta_{i,j}\}$ in l^2 such that the orthogonal projection $P : l^2 \to H_0$ satisfies $\|(I - P)u\| < a/2$.

Then

$$a \leqslant \langle y^i/\|y^i\|, u \rangle$$
$$= \langle Py^i/\|y^i\|, u \rangle + \langle y^i/\|y^i\|, (I-P)u \rangle$$
$$\leqslant \|Py^i\|/\|y^i\| + a/2,$$

and this is a contradiction because, in view of linear boundedness of K, $\lim_{i \to \infty} \|Py^i\|/\|y^i\| = 0$.

Now suppose $T: K \to K$ is nonexpansive with $\inf\{\|x - Tx\| : x \in K\} = b > 0$, and let $\|T0\| = d \geqslant b > 0$. Let $x(\alpha)$ denote the approximating curve $F_\alpha(0)$; thus for $\alpha \in [0, 1)$,

$$x(\alpha) = \alpha Tx(\alpha) = (\alpha/(1-\alpha))(Tx(\alpha) - x(\alpha))$$

and

$$Tx(\alpha) = (1-\alpha)^{-1}(Tx(\alpha) - x(\alpha)).$$

Hence

$$\|x(\alpha)\|^2 \geqslant \|Tx(\alpha) - T0\|^2$$
$$= \|Tx(\alpha)\|^2 - 2\langle Tx(\alpha), T0 \rangle + \|T0\|^2,$$

from which

$$2(1-\alpha)^{-1}\langle Tx(\alpha) - x(\alpha), T0 \rangle \geqslant \left[\left(\frac{1}{1-\alpha}\right)^2 - \left(\frac{\alpha}{1-\alpha}\right)^2\right] \|Tx(\alpha) - x(\alpha)\|^2 + \|T0\|^2.$$

Consequently

$$\left\langle \frac{Tx(\alpha) - x(\alpha)}{\|Tx(\alpha) - x(\alpha)\|}, \frac{T0}{\|T0\|} \right\rangle \geqslant \left(\frac{1+\alpha}{2}\right)(b/d) + \left(\frac{1-\alpha}{2}\right)(d/\|Tx(\alpha) - x(\alpha)\|),$$

and in turn this implies that the approximating curve $\{x(\alpha)\}$ lies, for example, in the bounded set $C \cap K$ where C is the cone

$$C = \left\{x : \langle x/\|x\|, T0/\|T0\| \rangle \geqslant \left(\frac{1+\alpha}{2}\right)(b/d)\right\}.$$

This contradicts the assumption $b > 0$.

The following simple question still remains unanswered. Does there exist an unbounded, closed, convex subset of a Banach space which has the fixed point property (f.p.p.) for nonexpansive mappings? If such a set exists it must of course be linearly bounded. The following example, due to R. Sine (1987), shows that no such set can exist in Hilbert space.

Example 12.4 Let K be an arbitrary unbounded, closed, convex set in a Hilbert space. If K is linearly unbounded, then K admits a nonexpansive translation $T: K \to K$ with Fix $T = \emptyset$. In general, a fixed-point free nonexpansive mapping $T: K \to K$ may be constructed as follows. Since K is unbounded, by the Uniform Boundedness Principle, there exists a linear functional f which is unbounded on K. Consider the sequence of nonempty, closed, convex sets

$$K_n = \{x \in K : f(x) \geqslant n\}, \qquad n = 1, 2, \ldots .$$

Let $P_n = \text{Proj}_{K_n}$ and let $\{c_n\}$ be a sequence of positive real numbers with $\Sigma_{n=1}^{\infty} c_n = 1$. Assume also that $\{c_n\}$ is chosen so that

$$\sum_{n=1}^{\infty} c_n \| P_n x \| < \infty \tag{12.9}$$

for some fixed $x \in K$. Then it routinely follows that (12.9) holds for all $x \in K$, and thus the mapping

$$T = \sum_{n=1}^{\infty} c_n P_n$$

is well defined and nonexpansive on K. If it were the case that T had a fixed point $w \in K$ then

$$w = \sum_{i=1}^{\infty} c_n P_n w$$

and

$$f(w) = \sum_{n=1}^{\infty} c_n f(P_n w). \tag{12.10}$$

But if $w \in K_n$ then $P_k w = w$ for $k \leqslant n$ while if $k > n$ and $w \notin K_k$, then $f(P_k w) > f(w)$. Clearly by (12.10) no such point of K exists.

One of the most important properties which distinguishes Hilbert spaces from other Banach spaces is the following well-known extension property first proved for euclidean spaces by Kirzbraun (1934) and extended to Hilbert spaces by Valentine (1943, 1945).

Theorem 12.4 (Kirzbraun–Valentine) *Let A be an arbitrary nonempty subset of a Hilbert space H and let $T: A \to H$ be nonexpansive. Then there exists a nonexpansive extension $\tilde{T}$ of T for which $\tilde{T}: H \to \overline{\text{conv}}\, T(A)$.*

Remark The above theorem remains valid for all lipschitzian mappings since, if $T: A \to H$ has Lipschitz constant k, the above version may be applied to the nonexpansive mapping $k^{-1} T$.

We now derive Theorem 12.4 from the following geometrical lemma proved later.

Lemma 12.2 *Let* $\{B(x_\alpha; r_\alpha)\}$ *and* $\{B(y_\alpha; r_\alpha)\}$, $\alpha \in \mathscr{A}$, *be two families of balls in* H *for which*

$$\| y_\alpha - y_\beta \| \leqslant \| x_\alpha - x_\beta \|, \qquad \alpha, \beta \in \mathscr{A}. \tag{12.11}$$

Then $\bigcap_{\alpha \in \mathscr{A}} B(x_\alpha; r_\alpha) \neq \emptyset \Rightarrow \bigcap_{\alpha \in \mathscr{A}} B(y_\alpha; r_\alpha) \neq \emptyset$.

Proof of Theorem 12.4 We may of course suppose $A \neq H$ and select $p \in H \backslash A$. Since $\| Tu - Tv \| \leqslant \| u - v \|$ for $u, v \in A$, we may apply the above lemma to the balls $\{B(x; \| x - p \|)\}$ and $\{B(Tx; \| x - p \|)\}$, $x \in A$, to conclude

$$C = \bigcap_{x \in A} B(Tx; \| x - p \|) \neq \emptyset.$$

Defining $\tilde{T}u = u$ if $u \in A$ and $\tilde{T}p \in C$ we obtain a nonexpansive extension U of T from A to $A \bigcup \{p\}$. A routine Zorn's Lemma argument yields a nonexpansive extension U of T defined on all of H. Now take $\tilde{T} = \text{Proj}_{\overline{\text{conv}} T(A)} \circ U$ to complete the proof.

We now derive Lemma 12.2 from its finite analogue:

Lemma 12.3 *Let* $\{B(x_i; r_i)\}$ *and* $\{B(y_i; r_i)\}$, $i = 1, \ldots, n$, *be two families of balls in* H *for which*

$$\| y_i - y_j \| \leqslant \| x_i - x_j \|, \qquad i, j = 1, \ldots, n. \tag{12.12}$$

Then $\bigcap_{i=1}^{n} B(x_i; r_i) \neq \emptyset \Rightarrow \bigcap_{i=1}^{n} B(y_i; r_i) \neq \emptyset$.

Proof of Lemma 12.2 Fix $\alpha_0 \in \mathscr{A}$. Then for each $\alpha \in \mathscr{A}$,

$$B(y_\alpha; r_\alpha) \cap B(y_{\alpha_0}; r_{\alpha_0})$$

is a closed and convex, hence weakly compact, subset of $B(y_{\alpha_0}; r_{\alpha_0})$. Therefore in order to complete the proof it suffices to show that the family $\{B(y_\alpha; r_\alpha) \cap B(y_{\alpha_0}; r_{\alpha_0})\}$ ($\alpha \in \mathscr{A}$) has the finite intersection property. So, fix $\{\alpha_1, \ldots, \alpha_n\} \subset \mathscr{A}$ and set $x_i = x_{\alpha_i}, y_i = y_{\alpha_i}, r_i = r_{\alpha_i}, i = 0, \ldots, n$. If $p \in \bigcap_{\alpha \in \mathscr{A}} B(x_\alpha; r_\alpha)$, then the subspace spanned by $\{p, x_0, \ldots, x_n, y_0, \ldots, y_n\}$ is a finite dimensional euclidean space and, moreover,

$$p \in \bigcap_{i=1}^{n} (B(x_i; r_i) \cap B(x_0; r_0)).$$

Thus by Lemma 12.3, $\bigcap_{i=1}^{n}(B(y_i; r_i) \bigcap B(y_0; r_0)) \neq \emptyset$, completing the proof.

Proof of Lemma 12.3 (Schoenberg, 1953) Assume $x \in \bigcap_{i=1}^{n} B(x_i; r_i)$. For each $\lambda \geqslant 0$ the set

$$P_\lambda = \{y \in X : \|y_i - y\| \leqslant \lambda \|x_i - x\|, i = 1, \ldots, n\}$$

is bounded, closed and convex and, for λ sufficiently large, obviously $P_\lambda \neq \emptyset$. Since $\mu \leqslant \lambda$ implies $P_\mu \subset P_\lambda$, by weak compactness there exists a smallest nonnegative number λ for which $P_\lambda \neq \emptyset$. The proof is complete if $\lambda \leqslant 1$, so we assume $\lambda > 1$.

Let $y \in P_\lambda$. Then we may assume, without loss of generality, that for some $k, 1 \leqslant k \leqslant n$,

$$\begin{aligned} \|y_i - y\| &> \|x_i - x\|, & i &= 1, \ldots, k; \\ \|y_i - y\| &\leqslant \|x_i - x\|, & i &= k+1, \ldots, n. \end{aligned} \tag{12.13}$$

We may also assume $y \in \mathrm{conv}\{y_1, \ldots, y_k\}$ since if this were not true it would be possible to contradict the minimality of λ by moving y very slightly in the direction perpendicular to any hyperplane separating y and $\{y_1, \ldots, y_k\}$ so as to decrease the distances $\|y_i - y\|, i = 1, \ldots, k$, while, at the same time, changing the distances $\|y_i - y\|, i = k+1, \ldots, n$ by an arbitrarily small amount. Therefore, suppose

$$y = \sum_{i=1}^{k} \mu_i y_i,$$

where $\mu_i \geqslant 0, i = 1, \ldots, k$ and $\Sigma_{i=1}^{k} \mu_i = 1$. Then for $i, j \in \{1, \ldots, k\}$,

$$\begin{aligned} \|x_i - x_j\|^2 &= \|x_i - x + x - x_j\|^2 \\ &= \|x_i - x\|^2 + \|x_j - x\|^2 - 2\langle x_i - x, x_j - x \rangle, \end{aligned}$$

and similarly

$$\|y_i - y_j\|^2 = \|y_i - y\|^2 + \|y_j - y\|^2 - 2\langle y_i - y, y_j - y \rangle.$$

In view of (12.12) and (12.13),

$$\langle y_i - y, y_j - y \rangle > \langle x_i - x, x_j - x \rangle$$

for $i, j \in \{1, \ldots, k\}$. Therefore

$$
\begin{aligned}
\|y - y\|^2 &= \left\| \sum_{i=1}^{k} \mu_i y_i - y \right\|^2 \\
&= \left\| \sum_{i=1}^{k} \mu_i (y_i - y) \right\|^2 \\
&= \sum_{i,j=1}^{k} \mu_i \mu_j \langle y_i - y, y_j - y \rangle \\
&> \sum_{i,j=1}^{k} \mu_i \mu_j \langle x_i - x, x_j - x \rangle \\
&= \left\| \sum_{i=1}^{k} \mu_i (x_i - x) \right\|^2,
\end{aligned}
$$

which is an obvious contradiction.

The following observation illustrates a simple example of an application of the Kirzbraun–Valentine Theorem.

Example 12.5 There exist nonconvex subsets A in a Hilbert space H which have the property that each $x \in \overline{\text{conv}}\, A$ has a unique nearest point $y \in A$. In other words, Proj_A is well defined for all $x \in \overline{\text{conv}}\, A$. Such sets are said to be Chebyshev with respect to their convex closures. (An example of such a set is the spherical slice

$$
S(y, a) = \{x \in H : \|x\| = 1, \langle y, x \rangle \geq a\}
$$

where $\|y\| = 1$ and $a \in (0, 1]$.)

Now suppose A is Chebyshev with respect to $K = \overline{\text{conv}}\, A$. Then A has the fixed point property for nonexpansive mappings. Indeed, let $T : A \to A$ be nonexpansive and let $\tilde{T}$ be an extension of T assured by Theorem 12.4. Then in particular $\tilde{T} : K \to K$ ($\tilde{T} : \overline{\text{conv}}\, A \to \overline{\text{conv}}\, T(A) \subset \overline{\text{conv}}\, A$) and so there exists $x \in K$ such that $\tilde{T}x = x$. If $x \in A$ then $\tilde{T}x = Tx = x$. On the other hand, if $x \notin A$, then for $y \in \text{Proj}_A x$, $\|Ty - x\| = \|\tilde{T}y - \tilde{T}x\| \leq \|y - x\|$, and this implies $Ty = y$. For further observations along this line, see Goebel and Massa, 1986; Goebel and Schöneberg, 1977.

13

Applications to accretivity

The study of nonexpansive mappings has been substantially motivated by the study of monotone and accretive operators, two classes of operators which arise naturally in the theory of differential equations.

The definitions given below have a very simple origin: if $D \subset \mathbb{R}$ and if $\varphi: D \to \mathbb{R}$, then φ is monotone increasing if for each $s, t \in D$,

$$(s-t)(\varphi(s) - \varphi(t)) \geqslant 0.$$

Let X be a Banach space with dual space X^* and let $D \subset X$. As before, we use the pairing $\langle x, j \rangle$ to denote $j(x)$, $x \in X, j \in X^*$.

Definition 13.1 A mapping $T: D \to X^*$ is said to be *monotone* if for each $u, v \in D$,

$$\langle u-v, Tu - Tv \rangle \geqslant 0. \tag{13.1}$$

A natural analogue of the above for mappings taking values in X is the following.

Definition 13.2 A mapping $T: D \to X$ is said to be *accretive* if for all $u, v \in D$ and some $j \in J(u-v)$,

$$\langle Tu - Tv, j \rangle \geqslant 0. \tag{13.2}$$

(Here J denotes the normalized duality mapping introduced earlier: For $x \in X$,

$$J(x) = \{ j \in X^*: \langle x, j \rangle = \|x\|^2 = \|j\|^2 \}.)$$

We note first that if X is a Hilbert space then $X = X^*$, the class of monotone and accretive operators defined in X coincide, and (13.1) denotes the usual inner product. Thus if X is specialized further to $X = \mathbb{R}$, (13.1) becomes $(u-v)(Tu - Tv) \geqslant 0$.

One connection between accretive operators and nonexpansive mappings is immediate. If $U: D \to X$ is nonexpansive, then for $T = I - U$, $x, y \in D$, and

136

$j \in J(x-y)$,

$$\langle Tx - Ty, j \rangle = \langle x - y - (Ux - Uy), j \rangle$$

$$= \|x-y\|^2 - \langle Ux - Uy, j \rangle$$

$$\geqslant \|x-y\|^2 - \|Ux - Uy\| \|x-y\|$$

$$\geqslant 0.$$

Thus T is accretive.

On the other hand, not all accretive mappings are of the form $I - U$ with U nonexpansive. A complete characterization of accretive mappings in metric terms has been given by Kato (1967).

Lemma 13.1 *Let X be a Banach space, $D \subset X$, and $T: D \rightarrow X$. Then T is accretive if and only if for each $x, y \in D$ and $\lambda \geqslant 0$,*

$$\|x-y\| \leqslant \|x-y + \lambda(Tx-Ty)\|.$$

Thus a mapping $T: D \rightarrow X$ is accretive if and only if the mapping $J_\lambda = (I + \lambda T)^{-1}$ (called the resolvent of T) is nonexpansive on its domain.

Using the above, it is possible to extend the definition of accretivity in a natural way to the multivalued case. For a given subset B of X, let $|B| = \inf\{\|x\| : x \in B\}$. Then a mapping $T: D \rightarrow 2^X (D \subset X)$ is said to be *accretive* if for each $x, y \in D$ and $\lambda \geqslant 0$,

$$\|x-y\| \leqslant |x-y + \lambda(Tx-Ty)|.$$

Again, the mapping $J_\lambda = (I + \lambda T)^{-1}$ is single-valued and nonexpansive on its domain. If it is the case that the domain of J_λ is all of X for some (hence all) $\lambda > 0$, then T is said to be *m-accretive* (sometimes called *hyperaccretive*).

The above observations not only reveal another connection between the accretive and nonexpansive mappings; they also lead in a natural way to several questions commonly asked in the theory: Under what conditions is an accretive mapping T *m*-accretive? When is an *m*-accretive mapping $T: D \rightarrow 2^X$ surjective in the sense that its range, $R(T)$, is all of X? Under what conditions does the equation $Tx = 0$ have a solution for an accretive mapping $T: D \rightarrow X$?

It is not surprising that fixed point theory for nonexpansive mappings plays a role in answering each of the above questions. In fact, if $T = I - U$ with U nonexpansive, each of the above questions can be reformulated as a problem in fixed point theory. We shall carry this full circle in the solvability case.

The theory of accretive operators is extensive. The solvability of many equations involving partial differential operators (laplacians) can be formulated as questions concerning accretive operators. We shall not discuss the details here but instead refer the reader to thorough treatments found elsewhere (e.g., Browder, 1976; Deimling, 1985; Martin, 1976). Our motivation for including this chapter is simply to note the usefulness of fixed point theory for nonexpansive mappings in another context and to illustrate the dependence of each theory upon the other.

We begin with the question of surjectivity. Let $T: D \to 2^X$ be an *m*-accretive mapping with resolvent J_λ, and let T_λ be the Yosida approximation: $T_\lambda = \lambda^{-1}(I - J_\lambda)$. We shall require the following well-known facts about these mappings (e.g., see Brezis, 1973; Crandall and Liggett, 1971; Crandall and Pazy, 1972). For $x \in D$:

(a) $\|T_\lambda x\| \leqslant |Tx|$ for all $\lambda \geqslant 0$;
(b) The mapping $\lambda \mapsto J_\lambda x$ is continuous for $\lambda \geqslant 0$, with $J_0 x = x$.

The following result holds for spaces satisfying the assumptions of all the fixed point theorems of Chapter 4. Several special cases of this theorem were established earlier by different methods (see, e.g., Browder, 1976; Kartsatos, 1978, 1981; Lange, 1973). This version is due to Kirk and Schöneberg (1980).

Theorem 13.1 *Let X be a Banach space for which the closed unit ball has the fixed point property (f.p.p.) for nonexpansive mappings, suppose $D \subset X$, and suppose $T: D \to 2^X$ is an m-accretive operator for which*

$$\lim_{\substack{\|x\| \to \infty \\ x \in D}} |Tx| = \infty. \tag{13.3}$$

Then $R(T) = X$.

Theorem 13.1 may be derived from the following. (Here ∂U denotes the boundary of U.)

Theorem 13.2 *Let X, D, and T be as in Theorem 13.1 and let $x_0 \in D$. Suppose U is a bounded neighborhood of x_0 and suppose*

$$|Tx_0| < |Tx| \qquad \text{for all } x \in \partial U \cap D. \tag{13.4}$$

Then there exists $y \in U$ such that $0 \in Ty$.

We shall require the following lemma (which holds in arbitrary spaces).

Lemma 13.2 *Under the assumptions of Theorem 13.2, $J_\lambda x_0 \in U$ for all $\lambda \geqslant 0$.*

Proof Since $\|T_\lambda x_0\| \leqslant |Tx_0|$ (by (a)) and since $T_\lambda x_0 \in TJ_\lambda x_0$ implies

$|TJ_\lambda x_0| \leqslant \|T_\lambda x_0\|$, it follows that $|TJ_\lambda x_0| \leqslant |Tx_0|$. Thus by (13.4), $J_\lambda x_0 \notin \partial U$. Also, $J_0 x_0 = x_0 \in U$, so (b) implies $J_\lambda x_0 \in U$ for all $\lambda \geqslant 0$.

Proof of Theorem 13.2 Since $J_\lambda x_0 \in U$ for all $\lambda \geqslant 0$ and since $|TJ_\lambda x_0| \leqslant \|T_\lambda x_0\| = \lambda^{-1}\|x_0 - J_\lambda x_0\|$, the fact that U is bounded implies

$$|TJ_\lambda x_0| \to 0 \qquad \text{as } \lambda \to \infty. \qquad (13.5)$$

Now let

$$d = \inf\{|Tx|: x \in \partial U \cap D\}.$$

If $0 \in Tx_0$ there is nothing to prove, so we suppose $0 \notin Tx_0$; hence $d > 0$. Also, in view of (13.5) we may suppose (redefining x_0 if necessary) that $|Tx_0| < d$. Since U is bounded it is possible to select $R > 0$ and $r > 0$ so that $U \subset B(x_0; R)$ and so that

$$|Tx_0| + 2r^{-1}R < d.$$

Fix $y \in B(x_0; R)$ and let $x = J_r y \in D$. We assert that $x \in U$. To see this, let $\bar{T}$ be the m-accretive operator defined by

$$\bar{T}x = Tx + r^{-1}(x_0 - y), \qquad x \in D,$$

and let $\bar{J}_r$ denote the resolvent of $\bar{T}$. Then for $z \in \partial U \cap D$,

$$\begin{aligned}
|\bar{T}z| &= |Tz + r^{-1}(x_0 - y)| \\
&\geqslant |Tx| - r^{-1}\|y - x_0\| \\
&\geqslant d - r^{-1}R \\
&> |Tx_0| + r^{-1}R \\
&\geqslant |Tx_0 + r^{-1}(x_0 - y)| \\
&= |\bar{T}x_0|.
\end{aligned}$$

Applying the lemma to $\bar{T}$ we conclude: $x = \bar{J}_r x_0 \in U$. The conclusion is now immediate. Since $U \subset B(x_0; R)$, J_r is a nonexpansive mapping of the ball $B(x_0; R)$ into itself. By assumption there exists $y \in B(x_0; R)$ such that $J_r y = y$; i.e., $y \in U \cap D$ and $0 \in Ty$.

Proof of Theorem 13.1 Theorem 13.1 is a direct corollary of the following:

Assertion If $|Tx_0| < r \leqslant \liminf\limits_{\substack{\|x\| \to \infty \\ x \in D}} |Tx|$, then $B(0; r) \subset R(T)$.

We first establish two preliminary facts.

Step 1 Under the assumption of the Assertion,

$$B(0; (\tfrac{1}{2})(r - |Tx_0|)) \subset R(T).$$

Proof Fix $y \in X$ such that $\| y \| < (\tfrac{1}{2})(r - |Tx_0|)$ and let $\bar{T} = T - y$. Then $\bar{T}$ is also *m*-accretive and it suffices to show $0 \in R(\bar{T})$. This will follow from Theorem 13.2 upon verification of the following.

Step 2 There exists $R > 0$ such that $|\bar{T}x_0| < |\bar{T}x|$ for $x \in D$ such that $\| x - x_0 \| = R$.

Proof If no such R exists then there exists a sequence $\{x_n\} \subset D$ such that $\| x_n \| \to \infty$ and $|\bar{T}x_n| \leqslant |\bar{T}x_0|$. It follows that $\| y \| \geqslant (\tfrac{1}{2})(|Tx_n| - |Tx_0|)$; hence $\| y \| \geqslant (\tfrac{1}{2})(r - |Tx_0|)$ — a contradiction.

Proof of the Assertion Fix $y \in X$ with $\| y \| < r$ and let

$$M = \{t \in [0, 1] : ty \in R(T)\}.$$

By Step 1, $M \neq \emptyset$. We show $1 \in M$. To accomplish this, let $t_0 = \sup M$ and select $t_n \in M$ with $t_n \to t_0$ as $n \to \infty$. Let $x_n \in D$ satisfy $t_n y \in Tx_n$, and for each n define T_n by

$$T_n x = T(x + x_n) - t_n y, \qquad x \in D(T_n) \equiv D - x_n.$$

Then clearly $0 \in T_n 0$. Also

$$\liminf_{\substack{\|x\| \to \infty \\ x \in D(T_n)}} |T_n x| \geqslant r - t_n \| y \|.$$

By applying Step 1 to the mapping T_n (replacing r with $r - t_n \| y \|$) we obtain

$$B(0; \tfrac{1}{2}(r - t_n \| y \|)) \subset R(T_n).$$

But if $t \geqslant t_0$ is sufficiently near t_0 it is possible to select n so that

$$(t - t_n) \| y \| < \tfrac{1}{2}(r - t_n \| y \|).$$

For such n there exists $z_n \in D - x_n$ such that $(t - t_n)y \in T_n z_n$, i.e., $ty \in T(z_n + x_n)$. This proves $t \in M$ and, in turn, $t_0 = 1 \in M$.

Further connections between nonexpansive mappings and accretive operators involve the so-called 'flow invariance' problem (cf., Reich, 1976b). Let D be a subset of a Banach space X and for $A : D \to X$ accretive consider the initial value problem:

$$du/dt = -Au; \qquad u(0) = x. \tag{13.6}$$

It turns out that if D is closed and if A is continuous, then the existence of a solution to (13.6) on $[0, \infty)$ depends, in a strong sense, only on whether or not the trajectory (local solution) remains in D. It is known that this is the case if the following 'flow invariance condition' is satisfied:

$$\lim_{h \to 0^+} h^{-1} \operatorname{dist}(x - hAx, D) = 0, \qquad x \in D. \tag{13.7}$$

Under the above assumptions the operator A generates a family of mappings $U_t : D \to D, t \in [0, \infty)$, by the relation

$$U_t x = u(t),$$

where $u(t)$ is the solution of (13.6). Then for each $s, t \in [0, \infty)$ and $x, y \in D$,

(a) $U_{s+t}(x) = U_s \circ U_t(x)$;
(b) $\|U_t(x) - U_t(y)\| \leqslant \|x - y\|$.

Thus the family $\{U_t\}_{t \geqslant 0}$ is a semigroup of nonexpansive mappings defined on D (with infinitesimal generator A).

The flow invariance condition (13.7) arises in related contexts, e.g., in the following theorem due to Martin which deals with the local solvability question.

Theorem 13.3 (Martin, 1973) *Let X be a Banach space with D a closed subset of X, and suppose $B: D \to X$ is continuous and strongly accretive in the sense that there exists $c > 0$ such that for each $u, v \in D$ there exists $j \in J(u - v)$ such that*

$$\langle Bu - Bv, j \rangle \geqslant c \|u - v\|^2.$$

Then if B satisfies (13.7), $0 \in B(D)$.

We now show that if D has the fixed point property (f.p.p.) for nonexpansive mappings, then the strong accretivity assumption in the above theorem may be relaxed. The approach is due independently to Reich and Ray; we follow Ray (1981).

Theorem 13.4 *Let X be a Banach space, let D be a closed, convex subset of X which has the f.p.p. for nonexpansive mappings, and let $A: D \to X$ be a continuous accretive operator which satisfies (13.7). Then $0 \in A(D)$.*

Proof Fix $t \in (0, 1)$, and note that by Kato's Lemma (Lemma 13.1) we have (taking $\lambda = t(1-t)^{-1}$):

$$\|((1-t)I + tA)(u) - ((1-t)I + tA)(v)\| = \|(1-t)((u-v) + t(Au - Av))\|$$

$$\geqslant (1-t)\|u-v\|, u, v \in D,$$

so the mapping $H = (1-t)((1-t)I + tA)^{-1}$ is nonexpansive on the set $((1-t)I + tA)(D)$. Setting $D_t = (1-t)D$, we show that $H: D_t \to D_t$.

For fixed $y \in D$, set $Bx = (1-t)(x-y) + tAx, x \in D$. Then, since A is accretive, if $u, v \in D$ there exists $j \in J(u-v)$ such that

$$\langle Bu - Bv, j \rangle = \langle (1-t)(u-v) + t(Au - Av), j \rangle$$

$$\geqslant (1-t)\|u-v\|^2.$$

Thus B satisfies the strong accretivity condition of Martin's Theorem. Next we show that B also satisfies (13.7). To see this, let $x \in D$ and $\epsilon > 0$. By (13.7), there exists $h_0 > 0$ such that if $h \in (0, h_0)$ then

$$\text{dist}(x - hAx, D) \leqslant (\tfrac{1}{2})h\epsilon.$$

For such h, there exists $u_h \in D$ such that

$$\|x - hAx - u_h\| \leqslant \text{dist}(x - hAx, D) + (\tfrac{1}{2})h\epsilon,$$

i.e.,

$$\|x - hAx - u_h\| \leqslant h\epsilon.$$

Therefore,

$$\lim_{h \to 0^+} h^{-1} \text{dist}(x - hBx, D) = \lim_{h \to 0^+} h^{-1} \text{dist}(t(x - hAx) + (1-t)(x - h(x-y)), D)$$

$$\leqslant \limsup_{h \to 0^+} h^{-1} \{t\|x - hAx - u_h\|$$

$$+ \text{dist}(tu_h + (1-t)(x - h(x-y)), D)\}.$$

Since $tu_h + (1-t)(x - h(x-y)) \in D$, it follows that

$$\lim_{h \to 0^+} h^{-1} \text{dist}(x - hBx, D) \leqslant \epsilon t,$$

and since $\epsilon > 0$ is arbitrary, the conclusion follows. Therefore by Martin's theorem, $0 \in B(D)$, i.e., there exists $x \in D$ such that

$$(1-t)y = (1-t)x + tAx.$$

Thus

$$H((1-t)y) = (1-t)((1-t)I - t)I + tA)^{-1}((1-t)y)$$

$$= (1-t)x.$$

Since $y \in D$ was arbitrary, this proves that $H: D_t \to D_t$ and, since D (hence D_t) has the f.p.p. for nonexpansive mappings, there exists a point $x_0 \in D_t$ such that $x_0 = Hx_0$. Taking $x_0 = (1-t)w$, we conclude

$$(1-t)w = (1-t)((1-t)I + tA)^{-1}((1-t)w)$$

from which $Aw = 0$ and the proof is complete.

It is interesting to note that Theorem 13.4 may itself be used to obtain fixed point theorems for nonexpansive mappings satisfying inwardness conditions. For a subset K of X and $x \in K$, the *inward set*, $I_K(x)$, of K with respect to x is the set

$$I_K(x) = \{x + c(u-x): u \in K, c \geqslant 1\}.$$

A mapping $T: K \to X$ is said to be *weakly inward* if Tx belongs to the closure of $I_K(x)$ for each $x \in K$.

The following is proved in Caristi, 1976

Lemma 13.3 *Let K be a convex subset of a Banach space and let $A: K \to X$. Then for $x \in K$,*

$$\lim_{h \to 0^+} h^{-1} \mathrm{dist}(x - hAx, K) = 0$$

if and only if the mapping $T = I - A$ is weakly inward on K.

We therefore have the following *fixed point result.*

Theorem 13.5 *Let X be a Banach space, let K be a closed and convex subset of X which has the f.p.p. with respect to nonexpansive mappings, and suppose $T: K \to X$ is nonexpansive and weakly inward on K. Then T has a fixed point in K.*

Proof As we noted at the beginning of this chapter, if $A = I - T$, then A is accretive on K. In view of Lemma 13.3 and Theorem 13.4, the conclusion follows.

14

Ultrafilter methods

The frequent use in previous chapters of such 'nongeometric' concepts as Zorn's Lemma in establishing the existence of fixed points for nonexpansive mappings serves to illustrate the limitations of a *purely* geometric theory and, in recent years, the development of the theory has been affected by more imaginative 'nonconstructive' approaches. Primary among these has been the introduction and use (by Maurey and others – Maurey, 1981; Lin, 1985; Khamsi, 1987; Belluce and Kirk, 1985) of model theoretic methods based on the concepts of ultrafilters and ultrapowers. These methods have produced a number of remarkably beautiful results which are deep and nonintuitive.

An excellent exposition on applications of the ultrapower concept in general Banach space theory is given in Heinrich (1980). Here we present an outline of the implications of this approach to geometric fixed point theory. Fundamentally important results arise almost immediately.

Let $\mathscr{T}$ be a nonempty set. A *filter* $\mathscr{F}$ in $\mathscr{T}$ is a collection of nonempty subsets of $\mathscr{T}$ satisfying

(a) $A, B \in \mathscr{F} \Rightarrow A \cap B \in \mathscr{F}$;
(b) $A \in \mathscr{F}$ and $B \supset A \Rightarrow B \in \mathscr{F}$.

The family of all filters on $\mathscr{T}$, ordered by set inclusion, satisfies the principle that each of its linearly ordered chains has an upper bound (the union of all filters in the family). Therefore, by Zorn's Lemma, every such filter is contained in a maximal filter (one not contained in any larger filter). Any filter maximal in this sense is called an *ultrafilter*. The simplest ultrafilters, and the only ones completely characterized, are those generated by a single element $i_0 \in \mathscr{T}$ (i.e., consisting of all subsets of $\mathscr{T}$ which contain i_0). These ultrafilters are called *trivial*, and all other ultrafilters are said to be *free ultrafilters*. Any ultrafilter which contains a sequence of sets $\mathscr{T}_1 \supset \mathscr{T}_2 \supset \cdots$ with $\bigcap_{k=1}^{\infty} \mathscr{T}_k = \varnothing$ is said to be *countably incomplete*, and obviously such an ultrafilter is free. (In fact, it might be noted that in the set $\mathbb{N}$ of natural numbers, an ultrafilter is countably incomplete if and only if it is free.)

Example 14.1 For the set $N = \{0, 1, \ldots\}$ the family of all sets containing a 'neighborhood of infinity', i.e., sets $A \subset \mathbb{N}$ such that $\mathbb{N} \backslash A$ is finite, is obviously a filter and thus is contained in the larger filter consisting of all sets $B \subset \mathbb{N}$ 'having density one' at infinity, i.e., sets B for which $\lim_{n \to \infty} (1/n) | \{ k \in B : k \leqslant n \}| = 1$, where $| \{ \ \} |$ denotes cardinality. Of course each of these filters is contained in an ultrafilter in $\mathbb{N}$.

The first fact we call attention to about ultrafilters is both trivial and fundamental.

Lemma 14.1 *If $\mathcal{U}$ is an ultrafilter in $\mathcal{T}$ and if $\mathcal{T}_0 \subset \mathcal{T}$, then either $\mathcal{T}_0 \in \mathcal{U}$ or $\mathcal{T} \backslash \mathcal{T}_0 \in \mathcal{U}$.*

Proof If neither $\mathcal{T}_0$ nor $\mathcal{T} \backslash \mathcal{T}_0$ is in $\mathcal{U}$, then for any $U \in \mathcal{U}$, $\mathcal{T}_0 \cap U$ is a nonempty proper subset of $\mathcal{T}_0$ which is not contained in $\mathcal{U}$. Thus the filter generated by $\mathcal{U}$ along with $U \cap \mathcal{T}_0$ properly contains $\mathcal{U}$ and contradicts the maximality of $\mathcal{U}$.

The usefulness of the ultrafilter concept in analysis is largely a result of the following definition of convergence.

Definition 14.1 Let S be a topological space and let $\mathcal{U}$ be an ultrafilter in a set $\mathcal{T}$. Let $x : \mathcal{T} \to S$, and for each $i \in \mathcal{T}$, denote $x(i) = x_i$. Then $\{ x_i : i \in \mathcal{T} \}$ is said to *converge with respect to $\mathcal{U}$* to an element $y \in S$ if for each neighborhood V of y the set $\{ i \in \mathcal{T} : x_i \in V \}$ is in $\mathcal{U}$. When this is the case we write:

$$\lim_{\mathcal{U}} x_i = y.$$

If an ultrafilter is not free and generated by $i_0 \in \mathcal{T}$, then it is always the case that $\lim_{\mathcal{U}} x_i = x_{i_0}$. However, with free ultrafilters the situation is much different. Convergence is common since such an ultrafilter consists of a 'rich' family of subsets. In fact, the following is true:

Theorem 14.1 *Let M be a compact Hausdorff space, let $\mathcal{U}$ be an ultrafilter in $\mathcal{T}$, and suppose $\{ x_i \}_{i \in \mathcal{T}} \subset M$. Then there exists a unique $y \in M$ such that*

$$\lim_{\mathcal{U}} x_i = y.$$

Proof Suppose $\{ x_i \}_{i \in \mathcal{T}}$ does not converge to any element of M. Then for each $x \in M$ there exists a neighborhood V_x of x for which

$$\mathcal{T}_x = \{ i \in \mathcal{T} : x_i \in V_x \} \notin \mathcal{U}.$$

Since M is compact, there exist $x_{j_1}, x_{j_2}, \ldots, x_{j_k} \in M$ for which $M \subset \bigcup_{k=1}^{n} V_{x_j}$. However, this implies

$$\bigcap_{j=1}^{n} (\mathcal{T} \setminus \mathcal{T}_{x_j}) = \emptyset$$

which is a contradiction since each set $\mathcal{T} \setminus \mathcal{T}_{x_j} \in \mathcal{U}$. The uniqueness of the limit is an easy consequence of the fact that M is Hausdorff.

We now turn to the construction of Banach space ultraproducts. As above, suppose $\mathcal{U}$ is an ultrafilter in $\mathcal{T}$, and for each $i \in \mathcal{T}$ let X_i be a Banach space. Let $l^{\infty}(\mathcal{T}, (X_i))$ denote the collection $\{x_i\}_{i \in \mathcal{T}}$ with $x_i \in X_i$ such that

$$\|\{x_i\}\| = \sup\{\|x_i\| : i \in \mathcal{T}\} < \infty.$$

Also let

$$\mathcal{N}(\mathcal{U}) = \{\{x_i\} \in l^{\infty}(\mathcal{T}, X_i) : \lim_{\mathcal{U}} \|x_i\| = 0\}.$$

It is easily verified that when equipped with the above norm $l^{\infty}(\mathcal{T}, X_i)$ is a Banach space and $\mathcal{N}(\mathcal{U})$ is a closed subspace of $l^{\infty}(\mathcal{T}, X_i)$.

Definition 14.2 With X_i and $\mathcal{U}$ as above, the *ultraproduct* $\{X_i\}_{\mathcal{U}}$ of the family $\{X_i, i \in \mathcal{T}\}$ is the quotient space

$$\{X_i\}_{\mathcal{U}} = l^{\infty}(\mathcal{T}, X_i)/\mathcal{N}(\mathcal{U})$$

equipped with the norm

$$\|\{x_i\}\| = \lim_{\mathcal{U}} \|x_i\|, \quad \{x_i\}_{\mathcal{U}} \in \{X_i\}_{\mathcal{U}}. \tag{14.1}$$

If $X_i \equiv X$ for $i \in \mathcal{T}$, then $\{X\}_{\mathcal{U}}$ is called the *ultrapower* of X with respect to $\mathcal{U}$. (Note that we use $\{x_i\}_{\mathcal{U}}$ to denote the equivalence class in $\{X_i\}_{\mathcal{U}}$ generated by $\{x_i\} \in l^{\infty}(\mathcal{T}, X_i)$.)

Remark 14.1 In view of Theorem 14.1, the limit (14.1) always exists and, moreover, it is easy to see that (14.1) is equivalent to the standard quotient norm.

An important related concept is that of the ultraproduct of operators. Let $\{X_i\}, \{Y_i\}, i \in \mathcal{T}$, be two families of Banach spaces, let $\{T_i\}$ be a family of linear operators such that $T_i : X_i \to Y_i$, and suppose the family $\{T_i\}$ is equibounded, i.e., suppose

$$\sup_{i\in \mathscr{I}} \| T_i \| < \infty.$$

Definition 14.3 With $\{X_i\}$, $\{Y_i\}$ and $\{T_i\}$ as above, the *ultraproduct* of the family $\{T_i\}$ is the operator $\{T_i\}_\mathscr{U} : \{X_i\}_\mathscr{U} \to \{Y_i\}_\mathscr{U}$ defined by

$$\{T_i\}_\mathscr{U} \{x_i\}_\mathscr{U} = \{T_i x_i\}_\mathscr{U}.$$

The above definition is consistent since $\lim_\mathscr{U} \|x_i\| = 0$ implies $\lim_\mathscr{U} \|Tx_i\| = 0$. Moreover,

$$\| \{T_i\}_\mathscr{U} \| = \lim_\mathscr{U} \| T_i \|.$$

Remark 14.2 Although the cardinality of $\mathscr{I}$ is relevant in certain model-theoretic considerations, for our purposes it will always suffice to take $\mathscr{U}$ to be a free ultrafilter in the set of natural numbers, *and we henceforth make this a standing assumption.*

Obviously, any ultrapower $\{X\}_\mathscr{U}$ of a Banach space X contains a subspace isometric to X (namely, the subspace consisting of all elements $\{x_i\}_\mathscr{U}$ where $x_i \equiv x \in X$). Although the ultrapower $\{X\}_\mathscr{U}$ is itself much 'richer' than X, its usefulness lies in its connection with X via the following theorem. (Recall (Chapter 5) that a Banach space Y is said to be *finitely representable* in X if for any finite dimensional subspace F_n of Y and any $\epsilon > 0$ there is a subspace E_n of X and an isomorphism $T : F_n \to E_n$ such that

$$(1+\epsilon)^{-1} \|y\| \leqslant \|Ty\| \leqslant (1+\epsilon)\|y\|$$

for all $y \in F_n$.)

Theorem 14.2 *The ultrapower $\{X\}_\mathscr{U}$ of a Banach space X is finitely representable in X.*

Proof Let M be a finite dimensional subspace of $\{X\}_\mathscr{U}$ and let $x^1, x^2, \ldots,$ $x^n \in M$ be an algebraic basis for M. For each $k \in \{1, \ldots, n\}$ select $x_i^k \in X$ such that $x^k = \{x_i^k\}_\mathscr{U}$ and let M_i be the subspace of X spanned by

$$\{x_i^k : k = 1, 2, \ldots, n\}.$$

Now define $T_i : M \to M_i$ by first setting

$$T_i x^k = x_i^k \qquad (k = 1, \ldots, n)$$

and then extending T_i linearly to all of M.

First note that the operators $\{T_i\}$ are equibounded (since

$$\|T_i\| \leqslant \sup\{\|x^k\|_{\mathcal{U}}/\|x^k\| : k = 1, \ldots, n\}).$$

Now suppose $x \in M$ is arbitrary, say

$$x = \sum_{k=1}^{n} \lambda_k x^k.$$

Then

$$\lim_{\mathcal{U}} \|T_i x\| = \lim_{\mathcal{U}} \left\| \sum_{k=1}^{n} \lambda_k x_i^k \right\|$$

$$= \left\| \sum_{k=1}^{n} \lambda_k \{x_i^k\}_{\mathcal{U}} \right\|$$

$$= \|x\|.$$

Therefore, for any $\epsilon > 0$ there exists a set $\mathcal{T}_x \in \mathcal{U}$ such that for all $i \in \mathcal{T}_x$,

$$\left(1 + \frac{\epsilon}{2}\right)^{-1} \|x\| \leqslant \|T_i x\| \leqslant \left(1 + \frac{\epsilon}{2}\right) \|x\|. \tag{14.2}$$

In turn, there exists a neighborhood V_x of x such that for all $y \in V_x$ and $i \in \mathcal{T}_x$,

$$(1 + \epsilon)^{-1} \|y\| \leqslant \|T_i y\| \leqslant (1 + \epsilon)\|y\|. \tag{14.3}$$

Applying this fact to the closed unit ball B in M (which is compact) it is possible to obtain a finite set $\{z_1, \ldots, z_p\} \subset B$ whose neighborhoods $\{V_{z_1}, \ldots, V_{z_p}\}$ cover B. Moreover, $\mathcal{T}_0 = \bigcap_{i=1}^{p} \mathcal{T}_{z_i} \in \mathcal{U}$. This, combined with (14.3), shows that M is $1 + \epsilon$ isomorphic to M_i for any $i \in \mathcal{T}_0$.

The above theorem has fundamental implications regarding the structure of the ultrapower $\{X\}_{\mathcal{U}}$ of X. The so-called 'local properties' of X, i.e., those properties governed by 'uniform behavior' of finite dimensional subspaces of X, are inherited by $\{X\}_{\mathcal{U}}$. For example, since the moduli of convexity and smoothness have 'two-dimensional character' (see Chapter 5),

$$\delta_X(\cdot) = \delta_{\{X_{\mathcal{U}}\}}(\cdot) \qquad \text{and} \qquad \rho_X(\cdot) = \rho_{\{X_{\mathcal{U}}\}}(\cdot);$$

consequently

$$\epsilon_0(X) = \epsilon_0(\{X\}_{\mathcal{U}}) \qquad \text{and} \qquad \rho'_X(0) = \rho'_{\{X\}_{\mathcal{U}}}(0).$$

For our first example of an application of the ultrapower technique we return to a fact we called attention to in Chapter 7.

Theorem 14.3 *All Banach spaces X satisfying $\rho'_X(0) < \frac{1}{2}$ have uniform normal structure.*

Proof We have already seen (Theorem 7.2) that X has normal structure. Suppose it is not uniform. Then there exists a sequence $\{K_n\}$ of bounded, closed and convex subsets of X, each containing 0 and having diameter 1, for which $\lim_{n\to\infty} r(K_n)=1$ (where $r(K_n)$ denotes the Chebyshev radius of K_n). Consider the set $K \subset \{X\}_\mathscr{U}$ defined by

$$K = \{K_i\}_\mathscr{U} = \{x \in \{X\}_\mathscr{U} : x = \{x_i\}_\mathscr{U}, x_i \in K_i\}$$

(where $\mathscr{U}$ is a free ultrafilter on $\mathbb{N}$). Then K is closed and convex in $\{X\}_\mathscr{U}$ with diam $K = 1$. Moreover, for any $x = \{x_i\}_\mathscr{U} \in K$ there exists $y = \{y_i\}_\mathscr{U} \in K$ such that

$$\lim_{i\to\infty} \|x_i - y_i\| = 1;$$

hence $\|x - y\|_\mathscr{U} = 1$ and K is diametral. On the other hand, since $\rho'_X(0) = \rho'_{\{X\}_\mathscr{U}}(0) < \tfrac{1}{2}$, $\{X\}_\mathscr{U}$ has normal structure, so we have a contradiction.

Further applications of ultrapower techniques to fixed point theory are based on the following construction.

Let K be a nonempty, convex, weakly compact set in X which is minimal invariant under the nonexpansive mapping T, and consider the set $\tilde{K} = \{K\}_\mathscr{U} \subset \{X\}_\mathscr{U}$ defined by

$$\tilde{K} = \{K\}_\mathscr{U} = \{x = \{x_i\}_\mathscr{U} \in \{X\}_\mathscr{U} : x_i \in K, i \in \mathbb{N}\}.$$

For simplicity, we assume diam $K = 1$ and $0 \in K$. Obviously $\tilde{K}$ contains a subset isometric to K; namely the one consisting of those elements of $\tilde{K}$ represented by constant sequences $\{x\}_\mathscr{U}$ with $x \in K$. We identify K with its natural isometric image in $\tilde{K}$ and note that the mapping $T: K \to K$ may be extended to a nonexpansive mapping $\tilde{T}: \tilde{K} \to \tilde{K}$ by setting

$$\tilde{T}x = \tilde{T}\{x_i\}_\mathscr{U} = \{Tx_i\}_\mathscr{U}. \tag{14.4}$$

Obviously diam $\tilde{K} = 1$ and $0 \in \tilde{K}$, but $\tilde{K}$ is *not* minimal invariant under $\tilde{T}$ (since it contains K).

Now, since K is bounded and convex it contains an approximate fixed point sequence $\{x_n\}$ for T. Set $\tilde{x} = \{x_n\}_\mathscr{U}$, and observe that

$$\|\tilde{T}\tilde{x} - \tilde{x}\| = \lim_\mathscr{U} \|Tx_n - x_n\| = \lim_{n\to\infty} \|Tx_n - x_n\| = 0.$$

This means that $\tilde{T}$ has a fixed point in $\tilde{K}$. In fact, the fixed points of $\tilde{T}$ are characterized as those points $\tilde{K}$ represented by sequences $\{x_n\}$ in K for which $\lim_\mathscr{U} \|Tx_n - x_n\| = 0$.

Now let $x = \{x\}_\mathscr{U} \in K$, and let $\tilde{x} = \{x_n\}_\mathscr{U}$ be a fixed point of $\tilde{T}$ represented by an approximate fixed point sequence $\{x_n\}$ of T. Then by Karlovitz's Lemma (Property 11.6 of Chapter 11),

$$\|x - \tilde{x}\| = \lim_{\mathcal{U}} \|x - x_n\| = \lim_{n \to \infty} \|x - x_n\| = 1.$$

On the other hand, if $\tilde{y} \in \text{Fix } \tilde{T}$ and $\tilde{y} = \{y_n\}_{\mathcal{U}}$, then some subsequence of $\{y_n\}$ must be an approximate fixed point sequence for T. It follows (from ultraproduct properties and the definition of the norm in $\{X\}_{\mathcal{U}}$) that for all $x \in K$,

$$\|x - \tilde{y}\| = 1.$$

Finally, using Karlovitz's Lemma and diagonalization, for any sequence $\{y_n\}$ in K there exists an approximate fixed point sequence $\{x_n\}$ for T in K such that $\lim_{n \to \infty} \|x_n - y_n\| = 1$.

We summarize the above facts as follows:

(a)   diam $K = $ diam $\tilde{K} = $ diam Fix $\tilde{T} = 1$.
(b) $K, \tilde{K}$ and Fix $\tilde{T}$ are diametral.
(c)   For any $x \in K$, dist$(x, \text{Fix } \tilde{T}) = 1$.
(d)   For any $\tilde{x} \in \text{Fix } \tilde{T}$, dist$(\tilde{x}, K) = 1$.
(e)   For any $x \in K$ and $\epsilon > 0$ there exists $\rho > 0$ such that $\|\tilde{y} - x\| > 1 - \epsilon$ whenever $\|\tilde{T}\tilde{y} - \tilde{y}\| < \rho$.

We also have the following:

Corollary 14.1 *If X does not have the fixed point property ($f.p.p.$) for weakly compact convex sets then the unit sphere in $\{X\}_{\mathcal{U}}$ contains a diametral convex set K with diam $K = 1$.*

The above observation is interesting, but so far it has not found applications because it is unclear how to identify spaces X for which $\{X\}_{\mathcal{U}}$ has nice rotundity properties. On the other hand, several authors have noted that certain geometrical properties of X may serve to contradict (a)–(e) and lead to the conclusion that any such minimal invariant set K must be a single point. We begin with an idea originating with Pei–Kee Lin (1985) concerning bases.

Let X be a Banach space with a basis $\{e^i\}_{i=1}^{\infty}$. Thus any element $x \in X$ has a unique representation

$$x = \sum_{i=1}^{\infty} \xi_i e^i, \tag{14.5}$$

and can be identified with the 'coordinate sequence' $\{\xi_i\}$. Let F be a subset of $\mathbb{N}$. The standard projection of X onto F is defined by

$$P_F x = \sum_{i \in F} \xi_i e^i,$$

and mappings of the form $2P_F - I$ are called *reflections* (in F).

Notice that in general the family of all standard projections (or reflections) is not equibounded. However, if all such mappings are equibounded, then the basis has two characteristic constants, c and λ, defined as follows:

$$c = \sup\{\|P_F\| : F \subset \mathbb{N}\};$$
$$\lambda = \sup\{\|2P_F - I\| : F \subset \mathbb{N}\}.$$

Note in particular that

$$\lambda = \sup\left\{ \left\| \sum_{i=1}^{\infty} \epsilon_i \xi_i e^i \right\| : \left\| \sum_{i=1}^{\infty} \xi_i e^i \right\| = 1, \epsilon_i = \pm 1 \right\}$$

$$= \sup\left\{ \left\| \sum_{i=1}^{n} \epsilon_i \xi_i e^i \right\| : \left\| \sum_{i=1}^{\infty} \xi_i e^i \right\| = 1, \epsilon_i = \pm 1, n = 1, 2, \ldots \right\}.$$

It is easy to verify that

$$1 \leqslant c \leqslant \lambda \leqslant 2c. \tag{14.6}$$

A basis $\{e^i\}$ for which $\lambda < +\infty$ is called *unconditional* and λ is called the *unconditional basis constant for* $\{e^i\}$ or, equivalently, $\{e^i\}$ is said to be *λ-unconditional*. In such a space, for any sequence $\{\epsilon_i\}$ with $\epsilon_i = \pm 1$,

$$\left\| \sum_{i=1}^{\infty} \epsilon_i \xi_i e^i \right\| \leqslant \lambda \left\| \sum_{i=1}^{\infty} \xi_i e^i \right\|$$

if and only if $\{\xi_i\}$ represents an element of X.

Example 14.2 The space c_0 and the classical l^p spaces, $1 \leqslant p < \infty$, all have a standard 1-unconditional basis $\{e^i\}$ given by $e^i = \{\delta_{ij}\}_{j=1}^{\infty}$. Also, Day's spaces D_1 and D_∞, and the space X_λ discussed in Examples 8.4 and 8.6, have standard 1-unconditional bases. Bynum's spaces $\Lambda_{p,q}$ of Example 6.2 have standard λ-unconditional bases for $\lambda = 2^{|p^{-1} - q^{-1}|}$.

The above examples serve to show that spaces possessing unconditional bases, or even 1-unconditional bases, need not be reflexive nor have normal structure. Nevertheless, Pei–Kee Lin (1985) proved the following:

Theorem 14.4 *Let X be a Banach space with a 1-unconditional basis $\{e^i\}$ and* *let K be a weakly compact convex subset of X. Then every nonexpansive mapping $T: K \to K$ has a fixed point.*

Proof We may assume K is minimal invariant under T. Let $\{x_n\}$ be an approximate fixed point sequence for T in K. If diam $K > 0$ we may assume, using translation and normalization, that

(a) diam $K = 1$;

(b) $0 \in K$;

(c) w-$\lim_{n \to \infty} x_n = 0$.

In view of Karlovitz's Lemma:

(d) $\lim_{n \to \infty} \|x_n\| = 1$,

and by passing to a subsequence we may further assume

(e) $\lim_{n \to \infty} \|x_n - x_{n+1}\| = 1$

or, even stronger,

(f) $\lim_{n \to \infty} \mathrm{dist}(x_n, \{x_{n+1}, x_{n+2},\ldots\}) = 1$.

Let $P_{[1,n]}$ denote the natural projection of X onto the subspace spanned by $\{e^1, \ldots, e^n\}$. Then by (c),

$$\lim_{n \to \infty} \|P_{[1,k]} x_n\| = 0$$

for each $k \in \mathbb{N}$. Also, for each $n \in \mathbb{N}$,

$$\lim_{k \to \infty} \|(I - P_{[1,k]}) x_n\| = 0.$$

By using the above and eventually passing to a subsequence it is possible to obtain a sequence $\{F_n\}$ of finite subsets of $\mathbb{N}$ with max $F_i \leqslant$ min F_{i+1} such that the natural projections $P_n \equiv P_{F_n}$ satisfy

(g) $\lim_{n \to \infty} \|P_n x_n\| = \lim_{n \to \infty} \|x_n\| = 1$;

(h) $\lim_{n \to \infty} \|(I - P_n) x_n\| = 0$;

and obviously,

(i) $P_n \circ P_m = 0$ if $n \neq m$.

We now translate the above observations into the terminology of ultrapowers. Consider the set $\tilde{K} = \{K\}_{\mathscr{U}} \subset \{X\}_{\mathscr{U}}$ and $\tilde{T}: \tilde{K} \to \tilde{K}$ defined as in (14.4). Observe that $\tilde{x} = \{x_n\}_{\mathscr{U}}$ and $\tilde{y} = \{x_{n+1}\}_{\mathscr{U}}$ are two fixed points of $\tilde{T}$, so by (e)

$$\tilde{x} = \tilde{T}\tilde{x}; \qquad \tilde{y} = \tilde{T}\tilde{y}; \qquad \|\tilde{x} - \tilde{y}\| = 1. \qquad (14.7)$$

Observe also that the mappings $\tilde{P} = \{P_n\}_{\mathcal{U}}$ and $\tilde{Q} = \{P_{n+1}\}_{\mathcal{U}}$ are defined in $\{X\}_{\mathcal{U}}$ and, by (g) and (h),

$$\tilde{P}\tilde{x} = \tilde{x}; \qquad \tilde{Q}\tilde{y} = \tilde{y}, \qquad (14.8)$$

while, since by (i) the projections $\{P_n\}$ are disjoint,

$$\tilde{P}\tilde{y} = 0; \qquad \tilde{Q}\tilde{x} = 0. \qquad (14.9)$$

Finally, for any point $x \in K \subset \tilde{K}$,

$$\tilde{P}x = 0; \qquad \tilde{Q}x = 0. \qquad (14.10)$$

We now invoke the fact that $\{e^i\}$ is a 1-unconditional basis for X. Since

$$\tilde{x} - \tilde{y} = \{x_n - x_{n+1}\}_{\mathcal{U}} = \{P_n x_n - P_{n+1} x_{n+1}\}_{\mathcal{U}},$$

it follows (from (e) and (h)) that

$$\|\tilde{x} - \tilde{y}\| = \lim_{\mathcal{U}} \|P_n x_n - P_{n+1} x_{n+1}\| = 1.$$

The fact that $\lambda = 1$ also implies that

$$\|\tilde{x} + \tilde{y}\| = \lim_{\mathcal{U}} \|P_n x_n - P_{n+1} x_{n+1}\|$$

$$= \lim_{n \to \infty} \|P_n x_n - P_{n+1} x_{n+1}\|$$

$$= 1.$$

Thus, finally,

$$\|\tilde{x} + \tilde{y}\| = \|\tilde{x} - \tilde{y}\| = 1. \qquad (14.11)$$

We now show that the above fact leads to a contradiction. Consider the set $\tilde{W} \subset \tilde{K}$ defined by

$$\tilde{W} = \{\tilde{w} \in \tilde{K}: \max\{\|\tilde{w} - x\|, \|\tilde{w} - \tilde{x}\|, \|\tilde{w} - \tilde{y}\|\} \leqslant \tfrac{1}{2} \text{ for some } x \in K\}.$$

Clearly $\tilde{W}$ is a closed and convex subset of $\tilde{K}$. Moreover, $\tilde{W}$ is nonempty since $\tilde{z} = \tfrac{1}{2}(\tilde{x} + \tilde{y})$ is, by (14.11), in $\tilde{W}$. Also, since $\tilde{x}$ and $\tilde{y}$ are fixed under $\tilde{T}$,

$$\max\{\|\tilde{T}\tilde{w} - Tx\|, \|\tilde{T}\tilde{w} - \tilde{x}\|, \|\tilde{T}\tilde{w} - \tilde{y}\|\} \leqslant \max\{\|\tilde{w} - x\|, \|\tilde{w} - \tilde{x}\|, \|\tilde{w} - \tilde{y}\|\},$$

showing that $\tilde{W}$ is invariant under $\tilde{T}$. We conclude therefore that $\tilde{W}$ contains an approximate fixed point sequence. On the other hand, for any point $\tilde{w} \in \tilde{W}$ there is a point $x \in K$ such that $\|\tilde{w} - x\| \leqslant \tfrac{1}{2}$. This leads to the following (which utilizes (14.7) and (14.10) in the last step)

$$\|\tilde{w}\| = \tfrac{1}{2}\|(\tilde{P} + \tilde{Q})\tilde{w} + (I - \tilde{P})\tilde{w} + (I - \tilde{Q})\tilde{w}\|$$
$$\leqslant \tfrac{1}{2}(\|(\tilde{P} + \tilde{Q})\tilde{w}\| + \|(I - P)\tilde{w}\| + \|(I - Q)\tilde{w}\|) \qquad (14.12)$$
$$\leqslant \tfrac{1}{2}(\|(\tilde{P} + \tilde{Q})(\tilde{w} - x)\| + \|(I - \tilde{P})(\tilde{w} - \tilde{x})\| + \|(I - Q)(\tilde{w} - \tilde{y})\|).$$

Since each of the mappings $\tilde{P} + \tilde{Q}, I - \tilde{P}, I - \tilde{Q}$ is represented as a sequence of projections and since the basis $\{e^i\}$ is 1-unconditional,

$$\|\tilde{P} + \tilde{Q}\| \leqslant 1, \qquad \|I - \tilde{P}\| \leqslant 1, \qquad \text{and } \|I - \tilde{Q}\| \leqslant 1.$$

Thus (14.12) yields

$$\|\tilde{w}\| \leqslant \tfrac{1}{2}(\|\tilde{w} - x\| + \|\tilde{w} - \tilde{x}\| + \|\tilde{w} - \tilde{y}\|) \leqslant \tfrac{3}{4}. \qquad (14.13)$$

But since $\tilde{W}$ contains an approximate fixed point sequence for $\tilde{T}$,

$$\sup\{\|\tilde{w}\| : \tilde{w} \in \tilde{W}\} = 1,$$

which contradicts (14.13).

Remark 14.3 The gap between $\tfrac{3}{4}$ and 1 in the final step of the above argument suggests that the result is not sharp. Actually, by sharpening the calculations Lin (1985), obtained the same result when X has a λ-unconditional basis for λ satisfying $6 - 3\lambda > \lambda^2$ (i.e., $\lambda < \tfrac{1}{2}(33^{1/2} - 3)$).

Other fixed point results for spaces with bases have been obtained using the ultrapower approach. For example P. K. Lin changed the assumption $\lambda = 1$ to $c = 1$ (thus $\lambda \leqslant 2$) and also obtained the following.

Theorem 14.5 (Lin, 1985) *If X is a superreflexive Banach space having an unconditional basis satisfying $c = 1$, then all weakly compact convex subsets of X have the f.p.p. for nonexpansive mappings.*

More recently, M. A. Khamsi has obtained a result which combines assumptions on λ and c.

Theorem 14.6 (Khamsi, 1987) *If X is a Banach space having an unconditional basis satisfying $c(\lambda + 2) < 4$, then all weakly compact, convex subsets of X have the f.p.p. for nonexpansive mappings.*

The question of course arises of whether the ultrapower approach is essential to the proofs of such theorems. In fact J. Elton, Pei–Kee Lin, E. Odell and S. Szarek (1983) have jointly obtained a (very technical) standard proof of the following deep and (in view of Alspach's example) surprising result of B. Maurey.

Theorem 14.7 (Maurey, 1981) *Bounded, closed, convex subsets of reflexive subspaces of $L^1[0, 1]$ have the f.p.p. for nonexpansive mappings.*

Maurey's proof of the above involves both the ultraproduct approach and delicate probabilistic techniques.

We state two additional results of Maurey's which are of interest. Both proofs use the ultrapower technique and the first combines this approach with classical results in complex function theory. We refer the reader to Elton *et al.* (1983) for a proof of the second.

Theorem 14.8 (Maurey, 1981) *Weakly compact, convex subsets of the Hardy space H^1 have the f.p.p. for nonexpansive mappings.*

Theorem 14.9 (Maurey, 1981) *Let K be a weakly compact, convex subset of a superreflexive space X and let $T: K \to K$ be an isometry (i.e., $\|Tx - Ty\| = \|x - y\|$, $x, y \in K$). Then T has a fixed point in K.*

There are other interesting technical results which illustrate the usefulness of the ultrafilter approach. For example Khamsi, in his Thèse de Doctorat (1987), (also see Khamsi and Thurpin, 1981) has shown that Karlovitz's Lemma holds for minimal weak* compact, convex sets in dual Banach spaces. His Thèse also contains a proof that weakly compact, convex subsets of the well-known James space J, which is nonreflexive yet isometrically isomorphic to its second dual J^{**}, have the f.p.p. for nonexpansive mappings.

The ultrafilter approach is not the only 'nonstandard' technique which has found application in this theory. For example, Bruck and Reich (e.g., 1981) have found the use of Banach Limits to be an expedient tool. Also, Kirk and others (e.g., Kirk, 1989; Kirk and Martinez, 1988; Kirk and Massa, 1987) have utilized the concept of an ultranet (which is related to an ultrafilter in the same sense that a net is related to a filter). This technique seems useful in situations where sequential approaches fail either because of insufficient geometric structure on the space or because of the 'nonsequential' nature of the topology under consideration. (Recall that we used nets in discussing Bruck's result on nonexpansive retracts (Theorem 11.3) because the topology under consideration was that of pointwise weak convergence, which is not metrizable.)

We conclude this chapter with a discussion of the ultranet approach, beginning with a review of the basic definitions and concepts regarding nets.

Definition 14.4 A set D is a *directed set* if there is a relation $\geqslant$ on D which satisfies

(D.1) $\alpha \geqslant \alpha$ for each $\alpha \in D$;
(D.2) if $\alpha \geqslant \beta$ and $\beta \geqslant \gamma$, then $\alpha \geqslant \gamma$,
(D.3) if $\alpha, \beta \in D$ there exists $\gamma \in D$ such that $\gamma \geqslant \alpha$ and $\gamma \geqslant \beta$.

The relation $\geqslant$ is a *linear order* on D if, in addition, either $\alpha \geqslant \beta$ or $\beta \geqslant \alpha$ for each $\alpha, \beta \in D$, with $\alpha = \beta$ when both occur.

Definition 14.5 A *net* (u, D) in a set S is a mapping $u: D \to S$ where D is a directed set. As with sequences, one frequently writes $u(\alpha) = s_\alpha$ and makes the identification $(u, D) = \{s_\alpha, \alpha \in D\}$. A net is said to be *eventually* in a set $G \subset S$ if there exists $\alpha_0 \in D$ such that $u(\alpha) \in G$ for all $\alpha \geqslant \alpha_0$. If S is a topological space, the net (u, D) is said to *converge* to $p \in S$ if (u, D) is eventually in each neighborhood of p. When this occurs we write $\lim_{\alpha \in D} s_\alpha = p$.

The notion of convergence defined above is the precise analogue of sequential convergence. In this respect, and in many others, sequences and nets are intuitively similar. However, there is one very important difference. If $\varphi: \mathbb{N} \to S$ is a sequence and if $f: \mathbb{N} \to \mathbb{N}$ satisfies $f(n_1) > f(n_2)$ whenever $n_1 > n_2$ in $\mathbb{N}$, then $\psi = \varphi \circ f: \mathbb{N} \to S$ is called a *subsequence* of φ. In the analogous definition for nets the mapping f is far less restricted.

Definition 14.6 If (u, D) is a net in S and if $f: E \to D$ is a mapping of a (possibly different) directed set E into D which satisfies

(N.1) for each $\alpha \in D$ there exists $e_0 \in E$ such that $f(e) \geqslant \alpha$ for all $e \geqslant e_0$,

then the net $(u \circ f, E)$ is called a *subnet* of (u, D).

Since subnets are defined by subordination, which is much less restrictive than inclusion, and since directed sets need not be linearly ordered, the above definition provides a very rich system of subnets for any net. In particular, it ensures the classical characterization of compact spaces: *A topological space S is compact if and only if each net in S has a subnet which converges to a point of S.*

The definition of subnet also leads to the abstract construction of ultranets. As we have seen, every filter is contained in a maximal filter called an ultrafilter. An analogous fact holds for nets. A net (u, D) in S is said to be an *ultranet* (also called a *universal* or *maximal* net) if given $G \subset S$, (u, D) is either eventually in G or eventually in the complement of G.

The usefulness of ultranets stems from the following facts.

(N.2) Every net has a subnet which is an ultranet.

(N.3) If (u, D) is an ultranet in S_1 and if $F: S_1 \to S_2$, then $(F \circ u, D)$ is an ultranet in S_2.

(N.4) If S is a compact topological space, then every ultranet in S converges.

The facts (N.3) and (N.4) are quite easy to verify, and a quick way to verify (N.2) is via the existence of ultrafilters (cf., e.g., Kelley, 1965). Let (u, D) be a net in S. It is easy to see that the family

$$\mathscr{F}_D = \{A \subset D : D \text{ is eventually in } A\}$$

is a filter. Thus $\mathscr{F}_D$ is contained in an ultrafilter $\mathscr{F}'_D$. Now let

$$D' = \{(d, F) : d \in F, F \in \mathscr{F}'_D\}.$$

Direct D' by setting $(e, E) \geqslant (d, F)$ provided $E \subset F$ and define $f: D' \to D$ by taking $f(d, F) = d$. Then for $A, E \in \mathscr{F}'_D$

$$f(e, E) \in A \text{ for each } e \in E \Leftrightarrow E \subset A \Leftrightarrow (e, E) \geqslant (a, A) \text{ for each } e \in E, a \in A.$$

Thus $\mathscr{F}'_D$ consists of precisely those sets $A \subset D$ such that the net $(u \circ f, D')$ is eventually in A. The fact that $(u \circ f, D')$ is an ultranet follows from the maximality of $\mathscr{F}'_D$. Also, since $A \in \mathscr{F}_D \Rightarrow A \in \mathscr{F}'_D$, the mapping f satisfies (N.1). Therefore $(u \circ f, D')$ is a subnet of (u, D).

To illustrate the usefulness of the ultranet concept in nonexpansive mapping theory we first define a transfinite iteration procedure.

As usual, let K be a weakly compact, convex subset of a Banach space and let $N \subset K^k$ denote the family of all nonexpansive mappings of K into K. (As we noted in Chapter 11, N is closed in K^K relative to the topology of weak pointwise convergence.)

The standard iteration for $T \in N$ is defined by induction, setting $T^0 = I$ and $T^{n+1} = T \circ T^n$, $n \in \mathbb{N}$. By using ultranets it is possible to extend this iteration and define T^α for any ordinal α. For our purposes it will suffice to consider only the set $\Omega = \Omega_0 \cup \{\omega_1\}$, where Ω_0 denotes the set of all countable ordinals and ω_1 denotes the first uncountable ordinal. We proceed by transfinite induction. Let $T^0 = I$, and suppose for $\alpha \in \Omega$ T^β has been defined for all $\beta < \alpha$. There are two possibilities; either α has an immediate predecessor $\bar{\beta}$ $(\alpha = \bar{\beta} + 1)$ or α is a limit ordinal. In the first case set $T^\alpha = T \circ T^\beta$. In the second case, in view of (N.2), it is possible to associate with $\alpha = \{\beta \in \Omega : \beta < \alpha\}$ a subnet (ξ_α, E_α) which is an ultranet. (For convenience we shall write $\xi_\alpha(\xi) = \alpha_\xi$ for $\xi \in E_\alpha$.) Then, in view of (N.4) and weak compactness of K, T^α can be defined by taking

$$T^\alpha x = w\text{-}\lim_{\xi \in E_\alpha} T^{\alpha_\xi} x, \qquad x \in K.$$

Therefore by the Principle of Transfinite Induction, the mappings T^α are well defined for all $\alpha \in \Omega$.

Remark 14.4 It should be noted that the subnet (ξ_α, E_α) of $\alpha \in \Omega$ used in defining T^α need not depend on the mapping T. Thus the iterates $\{T^\alpha\}$ may be defined uniformly for all mappings T – a fact used in the proof of Theorem 14.5.

Note in particular that each of the mappings T^α defined above is nonexpansive and that $T^\alpha x = x$ if $x \in$ Fix T. Therefore one would expect the iterations $\{T^\alpha x\}$ to approach Fix T (if Fix $T \neq \emptyset$) and, under certain circumstances, to become eventually constant. While not much is known in this direction, we present one theorem which has further consequences.

Recall that a Banach space X has Kadec–Klee *norm* (KK-norm; cf., Chapter 8) if for $\{x_n\}$ in X,

$$\left. \begin{array}{c} \|x_n\| \leqslant 1 \\ \mathrm{sep}\{x_n\} > 0 \\ w\text{-}\lim x_n = x \end{array} \right\} \Rightarrow \|x\| < 1.$$

This is a weak geometric condition which holds, for example, in all Δ-uniformly convex spaces. However it is strong enough to yield the following result, whose sequential analogue fails even in Hilbert space (Genel and Lindenstrauss, 1975).

Theorem 14.10 *Suppose X is a Banach space which is either strictly convex or has KK-norm, suppose K is a weakly compact, convex subset of X, and suppose $T: K \to K$ is nonexpansive with Fix $T \neq \emptyset$. Let $F = \frac{1}{2}(I + T)$ and let $\{F^\alpha\}$ be a transfinite iteration as defined above. Then for each $x \in K$, $F^\alpha x \in$ Fix T for some $\alpha \in \Omega_0$.*

Remark 14.5 Observe that once the transfinite iteration $\{T^\alpha x\}$ enters the set Fix T it becomes constant. Thus an immediate consequence of the above theorem is that $F^{\omega_1} : K \to$ Fix T is a nonexpansive retract of K onto Fix T. This provides an alternate proof of Bruck's Theorem (Theorem 11.3) for spaces X of the above type. (In fact, ultranets can also be used, in a different way, to prove Theorem 11.3 in its full generality.)

Let A be a subset of a Banach space X. An element $x \in A$ is called a *point of continuity* (PC) for A if the identity map I: $A \to A$ is continuous from the weak topology on A to the norm topology on A. If it is the case that $w\text{-}\lim_{n \to \infty} x_n = x$ implies $\lim_{n \to \infty} \|x_n\| = \|x\|$ for all $\{x_n\}$ in A, then x is said to be a point of sequential continuity (PSC) for A. In particular, X has a KK-norm if every point of the unit sphere $S(0; 1)$ of X is a (PSC) of $S(0; 1)$. It is shown in (Lin et al. 1989, Theorem 9) that if A is a closed, convex, weakly conditionally compact set in a Banach space then x is a (PSC) for A if and only if x is a (PC) for A. This immediately yields the following technical fact needed for our proof of Theorem 14.5.

Lemma 14.2 *With X and K as in Theorem 14.10, suppose A is a directed set and suppose $\{x_\alpha, \alpha \in A\}$ is a net in K which satisfies for some $p \in X$, $\|x_\alpha - p\| \equiv r > 0$ for all $\alpha \in A$. Then if $\{x_\alpha\}$ converges weakly but not strongly to $x \in K$, $\|x - p\| < r$.*

Proof of Theorem 14.10 Let $p \in \text{Fix } T$ and $x \in K$, and for $\alpha \in \Omega_0$ let $r_\alpha = \|p - F^\alpha x\|$. Since p is also a fixed point of F, $\{r_\alpha\}$ is a nonincreasing chain of real numbers with $\lim_\alpha r_\alpha = r \geq 0$. For each $n \in \mathbb{N}$ there exists $\alpha_n \in \Omega_0$ such that $\|p - F^{\alpha_n} x\| \leq r + 1/n$, and since the set $\{\alpha_1, \alpha_2, \ldots\}$ is countable, there exists $\alpha \in \Omega_0$ such that $\alpha \geq \alpha_n$ for all n. Clearly $\|p - F^\alpha x\| = r$, so we may suppose $r > 0$.

Now let γ denote the smallest limit ordinal larger than α. (Thus $\gamma = \alpha + \omega$ where ω is the first nonfinite ordinal.) By definition $F^\gamma x = w\text{-}\lim_{\xi \in E_\gamma} F^{\gamma\xi} x$ and, since there exists $\xi_0 \in E_\gamma$ such that $\xi \geq \xi_0 \Rightarrow \gamma_\xi \geq \alpha$, it is also the case that $F^\gamma x = w\text{-}\lim_{\xi \in E_\gamma / \xi \geq \xi_0} F^{\gamma\xi} x$. Since $\|p - F^\gamma x\| = \|p - F^{\gamma\xi} x\| = r$ for all $\xi \in E_\gamma$ with $\xi \geq \xi_0$, it follows from Lemma 14.2 that $F^\gamma x$ is the strong limit of $\{F^{\gamma\xi} x\}$. But since $\lim_{\xi \in E_\gamma / \xi \geq \xi_0} \|F^{\gamma\xi + 1} x - F^{\gamma\xi} x\| = 0$ by Theorem 9.4, we conclude that $F^\gamma x \in \text{Fix } F = \text{Fix } T$.

We now present an application of Theorem 14.10 to the study of nonexpansive mappings in product spaces.

Suppose $(X, \| \ \|_X)$ and $(Y, \| \ \|_Y)$ are Banach spaces, and let $X \oplus Y$ denote the product space of X and Y with l_2^∞ norm:

$$\|(x, y)\| = \max\{\|x\|_X, \|y\|_Y\}, \qquad x \in X, y \in Y.$$

It was shown in Kirk and Sternfeld (1984) that for X uniformly convex, if $K_1 \subset X$ is bounded, closed and convex and if $K_2 \subset Y$ is bounded and separable, then the assumption that K_2 has the f.p.p. for nonexpansive mappings assures that the same is true of $K_1 \oplus K_2$. By using Theorem 14.10 it is possible to generalize this result by simultaneously eliminating the separability assumption on K_2 and replacing the uniform convexity assumption on X with the much weaker assumption that X be reflexive and have KK-norm (see Kirk, 1988, cf., also Kirk and Martinez Yanez, 1988).

Theorem 14.11 *Let X and Y be Banach spaces and suppose X has KK-norm. Let $K_1 \subset X$ and $K_2 \subset Y$. Suppose K_1 is weakly compact and convex, and suppose both K_1 and K_2 have the f.p.p. for nonexpansive mappings. Then if $T: K_1 \oplus K_2 \to K_1 \otimes K_2$ is nonexpansive relative to the norm on $X \oplus Y$, T has a fixed point.*

Proof Let P_1 and P_2 denote, respectively, the natural coordinate projections of $X \oplus Y$ onto X and Y respectively, fix $y \in K_2$, and define $T_y: K_1 \to K_1$ by

$$T_y x = P_1 \circ T(x, y), \qquad x \in K_1.$$

It is trivial to check that T_y is nonexpansive. Let $F_y = (I + T_y)/2$ and fix $x_0 \in K_1$. By Theorem 14.10, for $\alpha \in \Omega_0$ sufficiently large $F_y^\alpha x_0 = p_y \in \text{Fix } T_y$. Thus

$$T_y(p_y) = p_y = P_1 \circ T(p_y, y).$$

Now let $u, v \in K_2$. Then

$$\begin{aligned}
\|F_u x_0 - F_v x_0\|_X &= \tfrac{1}{2}\|T_u x_0 - T_v x_0\|_X \\
&\leq \tfrac{1}{2}\|T(x_0, u) - T(x_0, v)\| \\
&\leq \|u - v\|_Y.
\end{aligned}$$

We make the inductive assumption that

$$\|F_u^\beta x_0 - F_v^\beta x_0\|_X \leq \|u - v\|_Y$$

for all $\beta < \alpha \in \Omega_0$. Then if α is a limit ordinal, $\|F_u^\alpha x_0 - F_v^\alpha x_0\|_X \leq \|u - v\|_Y$ by weak lower semicontinuity of the norm. On the other hand, if $\alpha = \alpha' + 1$, then

$$\begin{aligned}
\|F_u^\alpha x_0 - F_v^\alpha x_0\|_X &= \|F_u \circ F_u^{\alpha'} x_0 - F_v \circ F_v^{\alpha'} x_0\|_X \\
&\leq \tfrac{1}{2}\|T_u F_u^{\alpha'} x_0 - T_v F_v^{\alpha'} x_0\|_X + \tfrac{1}{2}\|F_u^{\alpha'} x_0 - F_v^{\alpha'} x_0\|_X \\
&= \tfrac{1}{2}\|P_1 \circ T(F_u^{\alpha'} x_0, u) - P_1 \circ T(F_v^{\alpha'} x_0, v)\|_X \\
&\quad + \tfrac{1}{2}\|F_u^{\alpha'} x_0 - F_v^{\alpha'} x_0\|_X \\
&\leq \tfrac{1}{2}\|T(F_u^{\alpha'} x_0, u) - T(F_v^{\alpha'} x_0, v)\| + \tfrac{1}{2}\|F_u^{\alpha'} x_0 - F_v^{\alpha'} x_0\|_X \\
&\leq \tfrac{1}{2}\max\{\|F_u^{\alpha'} x_0 - F_v^{\alpha'} x_0\|_X, \|u - v\|_Y\} + \tfrac{1}{2}\|F_u^{\alpha'} x_0 - F_v^{\alpha'} x_0\|_X \\
&\leq \|u - v\|_Y.
\end{aligned}$$

This completes the induction, so we conclude

$$\|F_u^\alpha x_0 - F_v^\alpha x_0\|_X \leqslant \|u - v\|_Y$$

for all $\alpha \in \Omega_0$. It follows that

$$\|p_u - p_v\|_X \leqslant \|u - v\|_Y, \qquad u, v \in K_2.$$

Now define $G: K_2 \to K_2$ by

$$Gy = P_2 \circ T(p_y, y), \qquad y \in K_2.$$

Then for $u, v \in K_2$,

$$\begin{aligned}
\|Gu - Gv\|_Y &= \|P_2 \circ T(p_u, u) - P_2 \circ T(p_v, v)\|_Y \\
&\leqslant \|T(p_u, u) - T(p_v, v)\| \\
&\leqslant \max\{\|p_u - p_v\|_X, \|u - v\|_Y\} \\
&= \|u - v\|_Y.
\end{aligned}$$

Therefore G is nonexpansive on K_2, so by assumption there exists $y \in K_2$ such that $Gy = y$. Thus

$$y = P_2 \circ T(p_y, y).$$

This combined with $p_y = P_1 \circ T(p_y, y)$ yields $T(p_y, y) = (p_y, y)$, completing the proof.

A detailed presentation of the methods, results and questions related to the nonstandard approaches in metric fixed point theory is well beyond the scope of this text. Moreover, results in this direction are still emerging and some (e.g., those involving ultranets) seem preliminary in nature. One would expect these nonstandard methods to provide material for an entirely separate treatment of the subject.

15

Set-valued mappings

The results concerning the existence of fixed points for contractions and nonexpansive mappings can be at least partially extended to the case of set-valued mappings. Recall (Example 2.4) that any complete metric space (M, ρ) is isometrically contained in the metric space $(\mathcal{M}, D)$ consisting of all nonempty, bounded and closed subsets of M. Here D denotes the Hausdorff metric defined for $A, B \in \mathcal{M}$ by

$$D(A, B) = \max\{d(A, B), d(B, A)\}$$

where

$$d(A, B) = \sup\{\text{dist}(a, B): a \in A\}.$$

As noted earlier, $(\mathcal{M}, D)$ is also a complete metric space.

By a set-valued mapping we mean a mapping of the form $T: M \to \mathcal{M}$; thus T assigns to each $x \in A$ a nonempty, bounded and closed subset of M. Such a mapping has Lipschitz constant $k > 0$ if

$$D(Tx, Tx) \leqslant k\rho(x, y), \qquad x, y \in M.$$

As before, T is called a *contraction mapping* if $k < 1$ and T is said to be *nonexpansive* if $k = 1$.

For set-valued mappings the notion of fixed point is modified. We say that $x \in M$ is a *fixed point of* T if and only if x is contained in its image, i.e., $x \in Tx$.

We begin with a simple generalization of Banach's Contraction Mapping Principle (see Markin 1968; Nadler, 1969).

Theorem 15.1. *Let (M, ρ) be a complete metric space and let $\mathcal{M}$ denote the collection of all nonempty, bounded, closed subsets of $\mathcal{M}$ endowed with the Hausdorff metric D. Suppose $T: (M, \rho) \to (\mathcal{M}, D)$ is a contraction mapping with Lipschitz constant $k < 1$. Then there exists $x \in M$ such that $x \in Tx$.*

In contrast with its point-valued counterpart, the fixed point x in Theorem 15.1 need not be unique. Indeed, if M is bounded then the mapping $Tx \equiv M$ for all $x \in M$ satisfies the assumptions of Theorem 15.1.

The key observation, used repeatedly in the proof of Theorem 15.1, is

162

a fact which follows immediately from the definition of D: If $X, Y \in \mathcal{M}$ and if $x \in X$, then for each $\epsilon > 0$ there exists $y \in Y$ such that

$$\rho(x, y) \leqslant D(X, Y) + \epsilon.$$

Proof of Theorem 15.1 Select $x_0 \in M$ and $x_1 \in Tx_0$. Then there exists $x_2 \in Tx_1$ such that

$$\rho(x_1, x_2) \leqslant D(Tx_0, Tx_1) + k,$$

and there exists $x_3 \in Tx_2$ such that

$$\rho(x_2, x_3) \leqslant D(Tx_1, Tx_2) + k^2.$$

In general, for each $i \in \mathbb{N}$, there exists $x_{i+1} \in Tx_i$ such that

$$\begin{aligned}
\rho(x_i, x_{i+1}) &\leqslant D(Tx_{i-1}, Tx_i) + k^i \\
&\leqslant k\rho(x_{i-1}, x_i) + k^i \\
&\leqslant k[D(Tx_{i-2}, Tx_i) + k^{i-1}] + k^i \\
&\leqslant k^2 \rho(x_{i-2}, x_{i-1}) + 2k^i \\
&\leqslant \cdots \\
&\leqslant k^i \rho(x_0, x_1) + ik^i.
\end{aligned}$$

Therefore

$$\sum_{i=0}^{\infty} \rho(x_i, x_{i+1}) \leqslant \rho(x_0, x_1) \left(\sum_{i=0}^{\infty} k^i \right) + \sum_{i=0}^{\infty} ik^i < \infty.$$

Hence $\{x_n\}$ is a Cauchy sequence, so there exists $x \in M$ such that $\lim_{n \to \infty} x_n = x$. Since T is continuous, $\lim_{n \to \infty} D(Tx_n, Tx) = 0$ and, since $x_n \in Tx_{n-1}$, $\lim_{n \to \infty} \mathrm{dist}(x_n, Tx) = 0$, whence $x \in Tx$ (since Tx is closed).

Before turning to the study of nonexpansive set-valued mappings we prove a stability result for set-valued contractions, due to Lim (1985), which has applications in other contexts (e.g., see Ricceri, 1987, 1988). Most of its proof is subsumed in the following:

Lemma 15.1 *Let M and $\mathcal{M}$ be as in Theorem 15.1 and suppose $T_i : M \to \mathcal{M}$, $i = 1, 2$, each have Lipschitz constant $k < 1$. Then if $F(T_1)$ and $F(T_2)$ denote the respective fixed point sets of T_1 and T_2,*

$$D(F(T_1), F(T_2)) \leqslant (1-k)^{-1} \sup_{x \in M} D(T_1 x, T_2 x).$$

Proof Let $\epsilon > 0$, choose $c > 0$, so that $c\sum_{n=1}^{\infty} nk^n < 1$, and set

$$\epsilon_1 = c\epsilon(1-k)^{-1}.$$

Select $x_0 \in F(T_1)$ and then select $x_1 \in T_2 x_0$ so that

$$\rho(x_0, x_1) \leqslant D(T_1 x_0, T_2 x_0) + \epsilon.$$

Since $D(T_2 x_1, T_2 x_0) \leqslant k \rho(x_1, x_0)$, it is possible to choose $x_2 \in T_2(x_1)$ so that $\rho(x_2, x_1) \leqslant k \rho(x_1, x_0) + k \epsilon_1$. Now define $\{x_n\}$ by induction so that $x_{n+1} \in T_2 x_n$ and

$$\rho(x_{n+1}, x_n) \leqslant k \rho(x_n, x_{n-1}) + k^n \epsilon_1, \qquad n = 1, 2, \ldots.$$

Then

$$\begin{aligned}
\rho(x_{n+1}, x_n) &\leqslant k \rho(x_n, x_{n-1}) + k^n \epsilon_1 \\
&\leqslant k(k \rho(x_{n-1}, x_{n-2}) + k^{n-1} \epsilon_1) + k^n \epsilon_1 \\
&= k^2 \rho(x_{n-1}, x_{n-2}) + 2 k^n \epsilon_1,
\end{aligned}$$

and continuing,

$$\rho(x_{n+1}, x_n) \leqslant k^n \rho(x_1, x_0) + n k^n \epsilon_1.$$

Therefore

$$\sum_{n=m}^{\infty} \rho(x_{n+1}, x_n) \leqslant k^m (1-k)^{-1} \rho(x_1, x_0) + \sum_{n=m}^{\infty} n k^n \epsilon_1.$$

Since the right side above tends to 0 as $m \to \infty$, $\{x_n\}$ is a Cauchy sequence with limit, say, $\bar{x}$. By continuity of T_2, $\lim_{n \to \infty} D(T_2 x_n, T_2 \bar{x}) = 0$ and, since $x_{n+1} \in T_2 x_n$, $\bar{x} \in T_2 \bar{x}$, i.e., $\bar{x} \in F(T_2)$. Furthermore,

$$\begin{aligned}
\rho(x_0, \bar{x}) &\leqslant \sum_{n=0}^{\infty} \rho(x_{n+1}, x_n) \\
&\leqslant (1-k)^{-1} \rho(x_1, x_0) + \sum_{n=1}^{\infty} n k^n \epsilon_1 \\
&\leqslant (1-k)^{-1} (\rho(x_1, x_0) + \epsilon) \\
&\leqslant (1-k)^{-1} (D(T_1 x_0, T_2 x_0) + 2\epsilon).
\end{aligned}$$

Reversing the roles of T_1 and T_2 we also conclude: For each $y_0 \in F(T_2)$ there exist $y_1 \in T_1 y_0$ and $\bar{y} \in F(T_1)$ such that

$$\rho(y_0, \bar{y}) \leqslant (1-k)^{-1} (D(T_1 y_0, T_2 y_0) + 2\epsilon).$$

Since $\epsilon > 0$ is arbitrary, the conclusion follows.

Theorem 15.2 *Let M and $\mathcal{M}$ be as in Theorem 15.1 and let $T_n: M \to \mathcal{M}$, $n = 0, 1, 2, \ldots$, be a sequence of contraction mappings each having Lipschitz constant $k < 1$. If $\lim_{n \to \infty} D(T_n x, T_0 x) = 0$ uniformly for $x \in M$, then $\lim_{n \to \infty} D(F(T_n), F(T_0)) = 0$.*

Proof Let $\epsilon > 0$ and choose N so that for $n \geqslant N$, $\sup_{x \in M} D(T_n x, T_0 x) < (1-k)\epsilon$. Then by Lemma 15.1 $D(F(T_n), F(T_0)) < \epsilon$ for such n.

We now turn to the study of set-valued, nonexpansive mappings in Banach spaces. Our basic setting is, as usual, a bounded, closed, convex subset K of a Banach space, and we shall assume that T is a nonexpansive mapping which assigns to each point $x \in K$ a nonempty, closed subset of K. We say in addition that T is *compact-valued* if Tx is compact for each $x \in K$ and *convex-valued* if Tx is convex for each $x \in K$.

Some facts from the theory of single-valued mappings remain unchanged. Under the above assumptions, if $p \in K$ and $\epsilon > 0$ are fixed, then the mapping T_ϵ defined by

$$T_\epsilon x = \epsilon p + (1-\epsilon) Tx, \qquad x \in K,$$

is a multivalued contraction mapping which approximates T for small ϵ; thus

$$\inf\{\operatorname{dist}(x, Tx): x \in K\} = 0.$$

In particular, T always has a fixed point if K is compact. Also, it is possible to use approximate fixed point sequences (sequences $\{x_n\}$ for which $\lim_{n \to \infty} \operatorname{dist}(x_n, Tx_n) = 0$) as a tool in much the same way as in the single-valued case.

The basic result in the nonexpansive theory is the following theorem, also due to Lim (1980).

Theorem 15.3 *Let X be a uniformly convex Banach space, let K be a bounded, closed and convex subset of X, and suppose T is a compact-valued, nonexpansive mapping defined on K with values contained in K. Then T has a fixed point.*

Earlier versions of Theorem 15.3 were obtained by Markin (1968) in a Hilbert space setting and by Browder (1976) for spaces possessing a weakly continuous duality mapping (Chapter 10). In both these instances T is assumed to be compact convex-valued. Another version of Theorem 15.3, due to Lami Dozo (1973), assumes X satisfies Opial's condition. For a survey of the literature as well as set-valued analogues of Theorem 13.5, see Downing and Ray, (1981).

The proof of Theorem 15.3 given here is patterned closely after that of Goebel (1975a) and Lim (1974b). It utilizes the asymptotic center technique and the following notion of regularity.

A bounded sequence $\{x_n\}$ in X is said to be *regular* relative to a subset

K of X if the asymptotic radii (relative to K) of all subsequences of $\{x_n\}$ are the same (i.e., $r(K, \{v_n\}) = r(K, \{x_n\})$ for each subsequence $\{v_n\}$ of $\{x_n\}$).

The following lemma will be used in the proof of Theorem 15.3 and in subsequent considerations.

Lemma 15.2 *Let X be a Banach space, K a subset of X, and $\{x_n\}$ a bounded sequence in X. Then $\{x_n\}$ has a subsequence which is regular relative to K.*

Proof For simplicity, we use the notation $\{v_n\} \prec \{u_n\}$ to indicate that $\{v_n\}$ is a subsequence of $\{u_n\}$, and we begin by setting

$$r_0 = \inf\{r(K, \{v_n\}): \{v_n\} \prec \{x_n\}\}.$$

Now select $\{v_n^1\} \prec \{x_n\}$ such that

$$r(K, \{v_n^1\}) < r_0 + 1,$$

and let

$$r_1 = \inf\{r(K, \{v_n\}): \{v_n\} \prec \{v_n^1\}\}.$$

Having defined $\{v_n^i\} \prec \{v_n^{i-1}\}$, set

$$r_i = \inf\{r(K, \{v_n\}): \{v_n\} \prec \{v_n^i\}\}$$

and select $\{v_n^{i+1}\} \prec \{v_n^i\}$ so that

$$r(K, \{v_n^{i+1}\}) \leqslant r_i + 1/(i+1). \tag{15.1}$$

Note that $r_1 \leqslant r_2 \leqslant r_3 \leqslant \cdots$; thus if $r = \lim_{r \to \infty} r_i$ then $\lim_{i \to \infty} r(K, \{v_n^{i+1}\}) = r$.

Now consider the diagonal sequence $\{v_k^k\}$ and let $\bar{r} = r(K, \{v_k^k\})$. Since $\{v_k^k\} \prec \{v_n^i\}$, clearly $\bar{r} \geqslant r_i$. On the other hand, since $\{v_k^k\} \prec \{v_n^{i+1}\}$, (15.1) implies $\bar{r} \leqslant r_i + 1/(i+1)$. Thus $\bar{r} = r$ and, since any subsequence $\{u_n\}$ of $\{v_k^k\}$ also satisfies $\{u_n\} \prec \{v_n^i\}$ and $\{u_n\} \prec \{v_n^{i+1}\}$ from which $r(K, \{u_n\}) = r$, we conclude $\{v_k^k\}$ is regular relative to K.

Proof of Theorem 15.3 In view of our preliminary observations and Lemma 15.2, there exists a sequence $\{x_n\}$ in K which satisfies

$$\lim_{n \to \infty} \text{dist}(x_n, Tx_n) = 0$$

and which is regular relative to K. Let

$$\{v\} = A(K, \{x_n\}) \qquad \text{and} \qquad r = r(K, \{x_n\}).$$

For each n, select $y_n \in Tx_n$ so that $\lim_{n \to \infty} \|y_n - x_n\| = 0$, and for each y_n select $z_n \in Tv$ so that

$$\|y_n - z_n\| \leqslant D(Tx_n, Tv) \leqslant \|x_n - v\|.$$

Now, since Tv is compact, some subsequence $\{z_{n_k}\}$ of $\{z_n\}$ converges to an element $w \in Tv$. Since

$$\|x_{n_k} - w\| \leqslant \|x_{n_k} - y_{n_k}\| + \|y_{n_k} - z_{n_k}\| + \|z_{n_k} - w\|$$

and

$$\|z_{n_k} - y_{n_k}\| \leqslant \|x_{n_k} - v\|,$$

it follows that

$$\limsup_{k \to \infty} \|w - x_{n_k}\| \leqslant r.$$

But since $\{x_n\}$ is regular, $v = w \in Tv$, completing the proof.

An analysis of the above proof reveals that uniform convexity of X, in its fullest sense, is not essential. It is enough to assume that the structure of K is such that all sequences $\{x_n\}$ in K have asymptotic centers consisting of precisely one point, and this may occur in more general situations (e.g., if K is weakly compact and X is uniformly convex in every direction). The conclusion of the proof is then based on the fact that in such a situation regular sequences actually share a much stronger property – all their subsequences have a common asymptotic center. We isolate this property as follows.

A sequence $\{x_n\}$ in X which is regular with respect to $K \subset X$ is said to be *asymptotically uniform* relative to K if $A(K, \{v_n\}) = A(K, \{x_n\})$ for every subsequence $\{v_n\}$ of $\{x_n\}$.

Note that it is always the case that $A(K, \{v_n\}) \supseteq A(K, \{x_n\})$ if $\{v_n\}$ is a subsequence of $\{x_n\}$, and strict inclusion may occur.

Example 15.1 Consider the space $\mathscr{C}[0, 1]$, and for $n = 0, 1, 2, \ldots$ and $k \in \{1, 2, \ldots, 2^n\}$ let $I_{n,k}$ denote the interval $[(k-1)/2^n, k/2^n)]$. Let $\{I_i\}$ be the sequence of intervals $\{I_{n,k}\}$ ordered lexicographically:

$$\{I_1, I_2, I_3, \ldots\} = \{I_{1,1}, I_{2,1}, I_{2,2}, I_{3,1}, I_{3,2}, I_{3,3}, \ldots\}.$$

Consider any sequence $\{\varphi_i\}$ of functions in $\mathscr{C}[0, 1]$ satisfying

(a) $\varphi_i(t) = 0$ for $t \notin I_i$;
(b) $\|\varphi_i\| = 1$ and φ_i takes on both the values 1 and -1 in I_i.

Then for any $x \in \mathscr{C}[0, 1]$,

$$r(x, \{\varphi_i\}) = 1 + \|x\| \geqslant 1$$

and for any subsequence $\{\psi_i\}$ of $\{\varphi_i\}$,

$$r(\mathscr{C}[0, 1], \{\psi_i\}) = r(\mathscr{C}[0, 1], \{\varphi_i\}) = 1.$$

Thus $\{\varphi_i\}$ is regular relative to $\mathscr{C}[0,1]$. However the asymptotic center of $\{\varphi_i\}$ relative to $\mathscr{C}[0,1]$ consists of only the zero function $\{0\}$, while subsequences of $\{\varphi_i\}$ may have larger asymptotic centers. For example, if $\{\psi_i\} = \{\varphi_{2i}\}$ then

$$A(\mathscr{C}[0,1], \{\psi_i\}) = \{x \in \mathscr{C}[0,1] : \|x\| \leqslant 1, x(0) = 0\}.$$

In spite of the above example, the following extension of Lemma 15.2 holds in all separable spaces.

Lemma 15.3 *Let X be a Banach space and let K be a separable subset of X. Then any bounded sequence $\{x_n\}$ in X contains an asymptotically uniform subsequence relative to K.*

Proof In view of Lemma 15.2 we may assume at the outset that $\{x_n\}$ is regular and, since K is separable, a routine diagonalization argument can be used to obtain a subsequence of $\{x_n\}$, which we again denote by $\{x_n\}$, such that $\lim_{n\to\infty} \|y - x_n\|$ exists for each $y \in K$. It is easy to see that such a sequence must be asymptotically uniform.

We now turn to a generalization of Theorem 15.3 initiated in Kirk (1986) and completed in Kirk and Massa 1988 using ultranets – an approach which might suggest that the purely metric approach has some limitations. We use a more straightforward approach here.

Suppose K is a bounded, closed, convex subset of a Banach space X, and assume that any sequence in K has a compact asymptotic center. Let T be a multivalued, nonexpansive mapping defined on K and having compact values lying in K. As in Theorem 15.3, it is possible to select an approximate fixed point sequence $\{x_n\}$ for T in K and it may further be assumed that $\{x_n\}$ is asymptotically regular. Moreover, for separable K $\{x_n\}$ can be assumed to be asymptotically uniform. In order to extend Theorem 15.3 to the nonseparable case we proceed as follows.

Let $V = A(K, \{x_n\})$, and define a sequence $\{K_n\}$ of subsets of K by taking

$$K_1 = \overline{\mathrm{conv}}(\{x_n\} \cup V),$$

$$K_{n+1} = \overline{\mathrm{conv}}(K_n \cup TK_n), \qquad n = 1, 2, \dots.$$

Now set

$$\tilde{K} = \overline{\bigcup_{n=1}^{\infty} K_n}$$

and observe that $\tilde{K}$ is a separable T-invariant subset of K which contains

$\{x_n\}$ and V. Thus

$$V = A(K, \{x_n\}) = A(\tilde{K}, \{x_n\})$$

and, in view of the regularity of $\{x_n\}$, for any subsequence $\{y_n\}$ of $\{x_n\}$,

$$V \subset A(\tilde{K}, \{y_n\}) = \tilde{K} \cap A(K, \{y_n\}).$$

Therefore any such subsequence $\{y_n\}$ has a nonempty, compact, asymptotic center with respect to $\tilde{K}$. Also, by Lemma 15.3, it is possible to select a subsequence $\{y_n\}$ of $\{x_n\}$ which is asymptotically uniform. In view of this, we may assume in what follows that $\{x_n\}$ was chosen *a priori* to be asymptotically uniform.

Now with $V = A(K, \{x_n\})$, select $v \in V$, and construct sequences $\{y_n\}$, $\{z_n\}$, and $\{z_{n_k}\}$, and the point $w \in Tv$ precisely as in the proof of Theorem 15.3. Note however, that since V is not necessarily a singleton it is not possible to conclude that v is a fixed point of T. On the other hand we do conclude that $Tv \cap V \neq \emptyset$. In particular a multivalued mapping F may be defined on the compact convex set V by taking

$$Fv = Tv \cap V, \qquad v \in V.$$

The mapping F need not be nonexpansive (nor even continuous), but it is easy to see that F is upper semicontinuous (i.e., for any sequence $\{u_n\}$ in V and any $v_n \in Fu_n$, the two conditions $\lim_{n \to \infty} u_n = u$ and $\lim_{n \to \infty} v_n = v$ imply $v \in Tu$). Although upper semicontinuity of F does not directly imply that F has a fixed point, classical results of Kakutani (1941) and Ky Fan (1960/61) assure the existence of such a point under the additional assumption that F has convex values. Finally, since F is convex-valued if T is, and since $Fv \subset Tv$, we have the following extension of Theorem 15.3.

Theorem 15.4 *Let X be a Banach space and let K be a bounded, closed, convex subset of X with the property that each sequence in K has a compact asymptotic center relative to K. Then any nonexpansive, compact, convex-valued mapping on K has a fixed point.*

We remark that the results of Kakutani and Fan alluded to above are of a 'nonmetric' nature and will be discussed in Chapter 18.

16

Uniformly lipschitzian mappings

While discussing stability properties of normal structure (Chapter 6) we observed that if $\|\cdot\|, \|\cdot\|_1$ are two equivalent norms on a Banach space X with

$$\alpha\|x\| \leqslant \|x\|_1 \leqslant \beta\|x\|, \qquad x \in X,$$

then the iterates of any nonexpansive mapping T defined on a subset D of $(X, \|\cdot\|_1)$ have uniformly bounded Lipschitz constants. Precisely,

$$\|Tx - Ty\|_1 \leqslant \|x - y\|_1, \qquad x, y \in D,$$

implies, for $k = \beta/\alpha$,

$$\|T^n x - T^n y\| \leqslant k\|x - y\|, \qquad x, y \in D, n = 1, 2, \ldots.$$

The same result holds if $\|\cdot\|_1$ is replaced with any metric ρ on D satisfying

$$\alpha\|x - y\| \leqslant \rho(x, y) \leqslant \beta\|x - y\|,$$

i.e., if $T: D \to X$ satisfies

$$\rho(Tx, Ty) \leqslant \rho(x, y), \qquad x, y \in D,$$

then for $k = \beta/\alpha$,

$$\|T^n x - T^n y\| \leqslant k\|x - y\|, \qquad x, y \in D, n = 1, 2, \ldots. \qquad (16.1)$$

Definition 16.1 A mapping $T: K \to K$ of a subset K of X is said to be *uniformly lipschitzian* if (16.1) holds for some $k > 0$.

Obviously all nonexpansive mappings are uniformly lipschitzian for $k = 1$, and the observation preceding the definition shows that all mappings which are nonexpansive with respect to some metric equivalent to the norm are uniformly lipschitzian. On the other hand, if $T: K \to K$ satisfies (16.1) then by setting

$$\rho(x, y) = \sup\{\|T^n x - T^n y\| : n = 0, 1, 2, \ldots\}, \qquad x, y \in K,$$

one obtains a metric ρ on K which is equivalent to the norm and relative to which T is nonexpansive:

$$\|x - y\| \leqslant \rho(x, y) \leqslant k\|x - y\|;$$
$$\rho(Tx, Ty) \leqslant \rho(x, y), \qquad x, y \in K.$$

Thus the class of uniformly lipschitzian mappings on K is completely characterized as the class of mappings on K which are nonexpansive relative to some metric on K which is equivalent to the norm.

It is natural to ask whether the fixed point property (f.p.p.) for nonexpansive mappings is stable in a sense stronger than that previously discussed; namely, if a set K has the (f.p.p.) for nonexpansive mappings, then does K also have the (f.p.p.) for mappings which are nonexpansive relative to a metric which is 'close' to the norm? The first positive result in this direction was the following.

Theorem 16.1 (Goebel and Kirk, 1973) *Let X be a uniformly convex Banach space with modulus of convexity δ and let K be a bounded closed and convex subset of X. Suppose $T: K \to K$ is uniformly lipschitzian relative to a constant k satisfying*

$$k(1 - \delta(1/k)) < 1. \tag{16.2}$$

Then T has a fixed point in K.

Note that the value of k in (16.2) is the same as that which arose in our discussion of the stability of normal structure (Theorem 6.3).

There are two questions connected with Theorem 16.1. First, is uniform convexity essential to the result? Second, is the estimate on k given in (16.2) sharp? The answer to both these is no. The original proof of Theorem 16.1 carries over easily to all spaces X for which $\epsilon_0(X) < 1$. Also, a sharper estimate for k has been obtained by E. A. Lifshitz in 1975 in which he generalizes Theorem 16.1 and even extends the result to a metric space setting. We take up the Lifshitz result next.

Let (M, ρ) be a complete metric space. The balls in M are said to be *c-regular* for $c \geqslant 1$ if the following holds: For any $k < c$ there are numbers $\mu, \alpha \in (0, 1)$ such that for any $x, y \in M$ and $r > 0$ with $\rho(x, y) \geqslant (1 - \mu)r$, the set $B(x; (1 + \mu)r) \cap B(y, k(1 + \mu)r)$ is contained in a closed ball of radius αr.

Two facts about the above definition are immediate: The balls in any metric space are 1-regular, and if the balls are c-regular for some $c > 1$ then they are c'-regular for all $c' \in [1, c]$.

Now set

$$\kappa(M) = \sup\{c \geqslant 1: \text{the balls in } M \text{ are } c\text{-regular}\}.$$

We shall refer to the number $\kappa(M)$ as the *Lifshitz character of M*. Obviously $\kappa(M) \geqslant 1$, and $\kappa(M)$ is the largest number for which M is $\kappa(M)$-regular.

For Banach spaces, the above definition has a simpler formulation: if X is a Banach space and if $r(D)$ denotes the Chebyshev radius of $D \subset X$, then

$$\kappa(X) = \sup\{c \geqslant 1 : r(B(0; 1) \cap B(x; c)) < 1 \text{ for all } \|x\| = 1\}.$$

It is not difficult to verify that $\kappa(H) = 2^{1/2}$ if H is a Hilbert space. Also $\kappa(X) > 1$ if and only if $\epsilon_0(X) < 1$ (see Downing and Turett, 1983).

Theorem 16.2 (Lifshitz, 1975) *Let (M, ρ) be a bounded and complete metric space, and suppose $T: M \to M$ is uniformly lipschitzian, with*

$$\rho(T^n x, T^n y) \leqslant k\rho(x, y), \qquad x, y \in M, n = 1, 2, \ldots,$$

where $k < \kappa(M)$. Then T has a fixed point in M.

Proof If $\kappa(M) = 1$ then T is a contraction mapping and there is nothing to prove. So, suppose $\kappa(M) > 1$. Select $x \in M$ and let $r(x)$ denote the radius of the smallest ball centered at x which contains an orbit of T, i.e., set

$$r(x) = \inf\{r > 0 : \text{for some } y \in M, \rho(x, T^n y) \leqslant r, n = 1, 2, \ldots\}.$$

Now let μ be the positive number associated with k in the definition of $\kappa(M)$-regular balls. Then given $x \in M$ there is an integer m such that

$$\rho(x, T^m x) \geqslant (1 - \mu)r(x),$$

and also there is a point $y \in M$ such that

$$\rho(x, T^n y) \leqslant (1 + \mu)r(x), \qquad n = 1, 2, \ldots.$$

Thus, by the definition of $\kappa(M)$-regularity, the set

$$D = B(x; (1 + \mu)r(x)) \cap B(T^m x; k(1 + \mu)r(x))$$

is contained in a closed ball centered say at $z \in M$, and having radius $\alpha r(x)$ where $\alpha > 1$. Next observe that for $n > m$,

$$\rho(T^m x, T^n y) \leqslant k\rho(x, T^{n-m} y) \leqslant k(1 + \mu)r(x),$$

and this shows that the orbit $\{T^n y : n > m\}$ is contained in D, hence in $B(z; \alpha r(x))$. This implies

$$r(z) \leqslant \alpha r(x).$$

Also, for any $u \in D$,

$$\rho(z, x) \leqslant \rho(z, u) + \rho(u, x)$$
$$\leqslant \alpha r(x) + (1 + \mu)r(x)$$
$$\leqslant A r(x)$$

where $A = \alpha + 1 + \mu$.

By setting $x_0 = x$ and $z(x_0) = z$, it is possible to construct a sequence $\{x_n\}$ with $x_{n+1} = z(x_n)$, where $z(x_n)$ is obtained via the above procedure. Thus $r(x_n) \leqslant \alpha^n r(x_0)$ and $\rho(x_{n+1}, x_n) \leqslant A r(x_n)$; hence $\{x_n\}$ converges to a fixed point of T.

The above theorem is of interest when $\kappa(M) > 1$, and this occurs in a Banach space X when $\epsilon_0(X) < 1$. In general, the inequality $k < \kappa(X)$ is less restrictive than the value of k given by (16.2). For example, in a Hilbert space H (16.2) is satisfied provided $k < 5^{1/2}/2$, while $\kappa(H) = 2^{1/2}$. Thus, in view of Lifshitz's Theorem we have the following.

Corollary 16.1 *Let K be a bounded, closed and convex subset of a Hilbert space and suppose $T: K \to K$ is uniformly lipschitzian with*

$$\| T^n x - T^n y \| \leqslant k \| x - y \|, \qquad x, y \in K, n = 1, 2, \ldots,$$

where $k < 2^{1/2}$. Then T has a fixed point in K.

We might remark that the central idea in the proof of Theorem 16.2 is similar to that of the original proof of Theorem 16.1, although the original is somewhat more constructive. There, the sequence $\{x_n\}$ is defined by simply taking x_{n+1} to be the asymptotic center of the orbit $\{T^k x_n : k = 1, 2, \ldots\}$ relative to K. As in the Lifschitz case, the sequence $\{x_n\}$ converges to a fixed point of T.

The above considerations raise several questions. Let X be a Banach space, let L denote the set of all numbers k for which any k-uniformly lipschitzian self-mapping of a bounded, closed, convex subset K of X has a fixed point, and set $\beta(X) = \sup L$. Obviously $\kappa(X) \leqslant \beta(X)$. Also, the family of k-uniformly lipschitzian mappings of K into K contains as a subfamily the collection of all mappings which are nonexpansive on K relative to an equivalent norm which is k-distant from the original norm ($k = \beta/\alpha$). Let $\gamma(X)$ denote the supremum of all numbers k for which mappings of this type have the fixed point property. Then clearly

$$1 \leqslant \kappa(X) \leqslant \beta(X) \leqslant \gamma(X).$$

Little is known about the precise nature of the above constants. Such questions as the following remain open. Is there a 'nice' space X, e.g., a space whose bounded, closed, convex sets have the f.p.p. for nonexpansive mappings, for which $\gamma(X) < \infty$? In general, are the above inequalities strict? However, as we shall prove in Chapter 19, it is known that in general $\beta(X) < \infty$. The following example demonstrates this fact in Hilbert space.

Example 16.1 (Baillon, 1975) Let $H = l^2$, and for points $x = (x_1, x_2, \ldots) \in l^2$ consider the three sets

$$B_1^+ = \{x \in l^2 : \|x\| \leqslant 1, x_1 = 0, x_j \geqslant 0 \text{ for } j = 2, 3, \ldots\};$$
$$S^+ = \{x \in l^2 : \|x\| = 1, x_j \geqslant 0 \text{ for } j = 1, 2, \ldots\};$$
$$S_1^+ = S^+ \cap B_1^+.$$

Let e denote the unit vector $(1, 0, 0, \ldots)$ and define $R : B_1^+ \to S_1^+$ by

$$Rx = \begin{cases} [\cos((\pi/2)\|x\|)]e + [\sin((\pi/2)\|x\|)](x/\|x\|) & \text{if } x \neq 0 \\ e & \text{if } x = 0. \end{cases}$$

(Note that R coincides with the identity on S_1^+.) For $x, y \in B_1^+$ with $x \neq y$, consider the curve

$$p(t) = R((1 - t)x + ty), \qquad 0 \leqslant t \leqslant 1.$$

A computation shows that $\|p'(t)\| \leqslant (\pi/2)\|x - y\|$ for all t. Thus

$$\|Rx - Ry\| \leqslant \int_0^1 \|p'(t)\| \, dt \leqslant (\pi/2)\|x - y\|.$$

Now let Q denote the shift operator defined by

$$Qx = Q(x_1, x_2, \ldots) = (0, x_1, x_2, \ldots),$$

and set $T = Q \circ R$. Then $T : B_1^+ \to S_1^+ \subset B_1^+$ and, since Q is an isometry with $T^n = Q^n \circ R$, T is k-uniformly lipschitzian with $k = \pi/2$. Also, since $x = Tx$ implies $\|x\| = 1$ and $x = Qx$, T is obviously fixed point free.

An earlier example of the above type, but for $k = 2$, may be found in Goebel, Kirk and Thele (1974). The general idea there is based on the observation that there exist bounded Q-shift invariant subsets K_0 of H which are nonconvex with $0 \notin K_0$, but for which there exists a lipschitzian retraction $R : \overline{\text{conv}} K_0 \to K_0$. The resulting mapping $T = Q \circ R$ is then uniformly lipschitzian on $K = \overline{\text{conv}} K_0$ with Fix $T = \emptyset$. However, in view of the fact that $k(H) = 2^{1/2}$, the uniform Lipschitz constant k for any such mapping (more precisely, for any retraction on a Q-invariant set which does not contain 0) must satisfy $k \geqslant 2^{1/2}$.

We conclude by observing that Baillon's example in conjunction with Lifschitz's Theorem implies

$$2^{1/2} \leqslant \beta(H) \leqslant \pi/2.$$

At present, the precise value of $\beta(H)$ remains unknown. It is known however that when restricted to the class of noncontractive mappings (mappings $T: K \to K$ for which $\|Tx - Ty\| \geqslant \|x - y\|$ for $x, y \in K$) the precise value of $\beta(H)$ is $\pi/2$ (see Tingley, 1984).

17

Rotative mappings

If $T: K \to K$ is nonexpansive for a subset K of a Banach space X, then for any $x \in K$ and $n \in \mathbb{N}$,

$$\|x - T^n x\| \leqslant \sum_{k=1}^{n} \|T^{k-1} x - T^k x\| \leqslant n \|x - Tx\|. \tag{17.1}$$

Equality occurs throughout (17.1) when the points $\{x, \ldots, T^n x\}$ all lie on a metric segment (a subset of X isometric with a real line interval) of length nd, where $d = \|x - Tx\|$. Thus if K is bounded, or even if T has bounded orbits, equality in (17.1) cannot hold for n sufficiently large if $Tx \neq x$ and, in fact, for such x the quantity

$$a_n(x) = \|x - T^n x\|/(n\|x - Tx\|) \tag{17.2}$$

converges to zero for all $x \in K$. Moreover, if T is a contraction mapping with Lipschitz constant $k < 1$, then

$$a_n(x) \leqslant (1/n)(1 + k + k^2 + \cdots + k^{n-1}) < 1. \tag{17.3}$$

It is generally the case, however, that the convergence of $\{a_n(x)\}$ is not uniform for $x \in K$, a fact which leads to the following definition.

Definition 17.1 A nonexpansive mapping $T: K \to K$ is said to be (a, n)-*rotative* for $a < n$ if for each $x \in K$,

$$\|x - T^n x\| \leqslant a \|x - Tx\|. \tag{17.4}$$

T is said to be *n-rotative* if it is (a, n)-rotative for some $a < n$, and *rotative* if it is *n*-rotative for some $n \in \mathbb{N}$.

Note that if a nonexpansive mapping is *n*-rotative then it is *m*-rotative for all $m \geqslant n$. Moreover, in view of (17.3), all contraction mappings are rotative (for all $n > 1$). However there are many other rotative mappings. Any periodic mapping T with period p (i.e., $T^p = I$) is $(0, p)$-rotative, and all rotations in the euclidean plane are rotative (a fact which inspired the use of the term).

If K is bounded, closed and convex, then any nonexpansive mapping $T: K \to K$ can be uniformly approximated by contraction mappings T_ϵ as $\epsilon \to 0$ by fixing $x_0 \in K$ and setting

$$T_\epsilon x = \epsilon x_0 + (1 - \epsilon) Tx, \qquad x \in K.$$

(Thus in topological terms, the family of rotative mappings, since it contains the contractions, is 'dense' in the family of nonexpansive mappings.)

The following theorem shows that the condition of rotativeness is actually quite strong.

Theorem 17.1 (Goebel and Koter, 1981a,b) *If K is a nonempty, closed and convex subset of a Banach space X, then any nonexpansive rotative mapping T: K→K has a fixed point. Moreover, the set Fix T is a nonexpansive retract of K.*

Remarks Note in particular that Theorem 17.1 does not require weak compactness or even boundedness of K, or any special geometric structure on K. Also, Theorem 17.1 implies that for any fixed point free, nonexpansive mapping the sequence $\{a_n\}$ of functions defined by (17.2) does not converge to zero uniformly on K. Since such a nonexpansive mapping cannot be rotative, for any $\epsilon > 0$ and $n \in \mathbb{N}$ there exists $x \in K$ such that

$$\|x - T^n x\| \geqslant (n - \epsilon)\|x - Tx\|.$$

(Therefore fixed point free, nonexpansive mappings always have orbits whose initial terms $\{x, \ldots, T^n x\}$ for large n are approximately linear.)

Proof of Theorem 17.1 Let $\alpha \in (0, 1)$ and consider the approximation curves (Chapter 11) given by:

$$F_\alpha(x) = (1 - \alpha)x + \alpha T F_\alpha x, \qquad x \in K. \tag{17.5}$$

By iterating F_α we obtain

$$F_\alpha^k x = (1 - \alpha)F_\alpha^{k-1} x + \alpha T F_\alpha^k x, \qquad k = 1, 2, \ldots. \tag{17.6}$$

Moreover, for any $x \in K$,

$$(1 - \alpha)(x - F_\alpha x) = \alpha(F_\alpha x - T F_\alpha x). \tag{17.7}$$

Now suppose T is (a, n)-rotative, i.e., suppose for all $x \in K$,

$$\|x - T^n x\| \leqslant a\|x - Tx\|. \tag{17.8}$$

Then

$$\|F_\alpha x - F_\alpha^2 x\| = \|F_\alpha x - (1 - \alpha)F_\alpha x - \alpha T F_\alpha^2 x\|$$

$$= \alpha\|F_\alpha x - T F_\alpha^2 x\|$$

$$\leqslant \alpha\|F_\alpha x - T^n F_\alpha x\| + \alpha\|T^n F_\alpha x - T F_\alpha^2 x\|$$

$$\leqslant \alpha a\|F_\alpha x - T F_\alpha x\| + \alpha\|T^{n-1} F_\alpha x - F_\alpha^2 x\|$$

$$= (1 - \alpha)a\|x - F_\alpha x\| + \alpha\|T^{n-1} F_\alpha x - F_\alpha^2 x\|;$$

hence for all $x \in K$,

$$\|F_\alpha x - F_\alpha^2 x\| \leqslant (1-\alpha)a\|x - F_\alpha x\| + \alpha\|T^{n-1}F_\alpha x - F_\alpha^2 x\|. \quad (17.9)$$

Note that in obtaining (17.9) we used the rotativeness condition (17.8). Now, using only the fact that T is nonexpansive, we proceed by induction to establish the following.

$$\alpha\|T^{m-1}F_\alpha x - F_\alpha^2 x\| \leqslant [(m-1) - m\alpha + \alpha^m]\|x - F_\alpha x\|$$
$$+ \alpha^m\|F_\alpha x - F_\alpha^2 x\|, \qquad x \in K, m \geqslant 2. \quad (17.10)$$

Indeed, for $m = 2$,

$$\alpha\|TF_\alpha x - F_\alpha^2 x\| = \alpha\|TF_\alpha x - (1-\alpha)F_\alpha x - \alpha TF_\alpha^2 x - \alpha TF_\alpha x + \alpha TF_\alpha x\|$$
$$\leqslant \alpha(1-\alpha)\|TF_\alpha x - F_\alpha x\| + \alpha^2\|TF_\alpha^2 x - TF_\alpha x\|$$
$$\leqslant (1-\alpha)^2\|x - F_\alpha x\| + \alpha^2\|F_\alpha^2 x - F_\alpha x\|.$$

Now suppose (17.10) holds for $m = j$, and consider $m = j + 1$:

$$\alpha\|T^j F_\alpha x - F_\alpha^2 x\| = \alpha\|(1-\alpha)T^j F_\alpha x + \alpha T^j F_\alpha x - (1-\alpha)F_\alpha x - \alpha TF_\alpha^2 x\|$$
$$\leqslant \alpha(1-\alpha)\|T^j F_\alpha x - F_\alpha x\| + \alpha^2\|T^j F_\alpha x - TF_\alpha^2 x\|$$
$$\leqslant j\alpha(1-\alpha)\|F_\alpha x - TF_\alpha x\| + \alpha^2\|T^{j-1}F_\alpha x - F_\alpha^2 x\|$$
$$= j(1-\alpha)^2\|x - F_\alpha x\| + \alpha^2\|T^{j-1}F_\alpha x - F_\alpha^2 x\|,$$

so by the inductive hypothesis

$$\alpha\|T^j F_\alpha x - F_\alpha^2 x\| \leqslant j(1-\alpha)^2\|x - F_\alpha x\|$$
$$+ \alpha[(j-1) - j\alpha + \alpha^j]\|x - F_\alpha x\| + \alpha^{j+1}\|F_\alpha x - F_\alpha^2 x\|$$
$$= [j - (j+1)\alpha + \alpha^{j+1}]\|x - F_\alpha x\| + \alpha^{j+1}\|F_\alpha x - F_\alpha^2 x\|.$$

This establishes (17.10).

Finally, using (17.9) and (17.10) we conclude

$$\|F_\alpha x - F_\alpha^2 x\| \leqslant (1-\alpha^n)^{-1}[(1-\alpha)a + (n-1) - n\alpha + \alpha^n]\|x - F_\alpha x\|$$
$$= \left[(a+n)\left(\sum_{i=0}^{n-1}\alpha^i\right)^{-1} - 1\right]\|x - F_\alpha x\|$$
$$= g(\alpha)\|x - F_\alpha x\|, \qquad x \in K. \quad (17.11)$$

Since g is continuous and decreasing for $\alpha \in (0, 1]$ with $g(1) = a/n < 1$, there exists $b \in (0, 1)$ such that $g(\alpha) < 1$ for $\alpha \in (b, 1)$. For such α, the iterates $\{F_\alpha^n x\}$ converge for any $x \in K$ and, since Fix $F_\alpha =$ Fix T, these iterates converge to a fixed point of T. Moreover, the mapping $R: K \to \text{Fix } T$ given by

$$Rx = \lim_{n \to \infty} F_\alpha^n x$$

is a nonexpansive retraction.

The rotativeness condition (17.4) is actually independent of the nonexpansiveness of T (and, in fact, this condition may be satisfied by very irregular mappings). Taking this into account, an analysis of the proof of Theorem 17.1 shows room for improvement.

Definition 17.2 Denote by $\Phi(a, n, k, K)$ the class of all mappings $T: K \to K$ satisfying for $x, y \in K$,

$$\|Tx - Ty\| \leqslant k\|x - y\|; \tag{17.12}$$

$$\|x - T^n x\| \leqslant a\|x - Tx\|. \tag{17.13}$$

A mapping $T \in \Phi(a, n, k, K)$ is said to be *k-lipschitzian (a, n)-rotative on K*.

It turns out that even k-lipschitzian (a, n)-rotative mappings have fixed points provided $k > 1$ is not too large.

Definition 17.3 For a Banach space X and any pair (a, n), $a < n$ set, for $K \subset X$,

$$\gamma(a, n, K, X) = \sup\{\gamma: T \in \Phi(a, n, k, K) \text{ for } k < \gamma \text{ implies Fix } T \neq \emptyset\},$$

and let

$$\gamma(a, n, X) = \inf\{\gamma(a, n, K, X): K \text{ a bounded, closed convex subset of } X\}.$$

Theorem 17.2 *For any Banach space X and any pair (a, n) with $a < n$, $\gamma(a, n, X) > 1$.*

Proof This proof is patterned closely after that of Theorem 17.1. Let $T \in \Phi(a, n, k, K)$. Since T is k-lipschitzian for any $x \in K$ the Banach Contraction Principle implies that the equation

$$y = (1 - \alpha)x + \alpha Ty$$

has a unique solution $F_\alpha x$ if $\alpha \in [0, 1/k)$.

Thus, fixing such α,

$$F_\alpha x = (1 - \alpha)x + \alpha T F_\alpha x \tag{17.14}$$

and consequently,

$$\|F_\alpha x - F_\alpha y\| \leqslant \frac{1-\alpha}{1-k\alpha}\|x-y\|,$$

which shows that F_α is lipschitzian.

Now, using the fact that (17.7) remains valid, proceed exactly as in the proof of Theorem 17.1 but use (17.12) instead of nonexpansiveness to obtain the following counterparts of (17.9), (17.10), (17.11):

$$\|F_\alpha x - F_\alpha^2 x\| \leqslant (1-\alpha)a\|x - F_\alpha x\| + \alpha k\|T^{n-1}F_\alpha x - F_\alpha^2 x\|; \quad (17.9')$$

$$\alpha k\|T^{m-1}F_\alpha x - F_\alpha^2 x\| \leqslant (1-\alpha)\left(\frac{k^m-1}{k-1} - \frac{1-\alpha^m k^m}{1-\alpha k}\right)\|x - F_\alpha x\|$$
$$+ \alpha^m k^m\|F_\alpha x - F_\alpha^2 x\|; \quad (17.10')$$

$$\|F_\alpha x - F_\alpha^2 x\| \leqslant \frac{1-\alpha}{1-\alpha k}\left[\frac{a+\dfrac{k^n-1}{k-1}}{\displaystyle\sum_{i=0}^{n-1}(\alpha k)^i} - 1\right]\|x - F_\alpha x\|$$

$$= g(\alpha, k)\|x - F_\alpha x\|. \quad (17.11')$$

Note that $\lim_{k\to 1} g(\alpha, k) = g(\alpha)$ as defined in (17.11). Thus, since $g(\alpha) < 1$ for α sufficiently near 1, $g(\alpha, k) < 1$ if both α and k are sufficiently near 1 (but with $k > 1$). For such α, k, the iterates $\{F_\alpha^n x\}$ for any $x \in K$ converge to a point of Fix $F_\alpha = $ Fix T.

Theorem 17.2 is a qualitative result which raises a number of technical yet attractive questions concerning the precise values of $\gamma(a, n, X)$. Even the exact value of $\gamma(0, n, X)$ is of interest since it characterizes the fixed point behavior of mappings of period n. Preliminary investigations in the periodic case were initiated by Goebel and Zlotkiewicz (Goebel, 1973; Goebel and Zlotkiewicz, 1971) and by Kirk (1971). In particular, the first two papers deal with involutions (mappings T for which $T^2 = I$). It is shown that such mappings always have fixed points if they are k-lipschitzian for $k < 2$ and moreover, if the space X satisfies $\epsilon_0(X) < 1$, the same is true for k-lipschitzian involutions where k satisfies

$$\left(\frac{k}{2}\right)\left(1 - \delta_X\left(\frac{2}{k}\right)\right) < 1. \quad (17.15)$$

The proofs of these facts are straightforward verifications that starting from any $x \in K$, the sequence of iterates $\{F^n x\}$ for $F = \frac{1}{2}(I + T)$ always converges to a fixed point of T. (We remark that these estimates on k are twice as large as the constant k arising in the study of stability of normal structure (Chapter 6).) Thus, in our notation, $\gamma(0, 2, X) \geqslant 2k_0$ where k_0 is the unique solution of the equation

$$k_0(1 - \delta_X(1/k_0)) = 1.$$

This implies $\gamma(0, 2, X) \geqslant 2$ for any space X while, in particular, $\gamma(0, 2, H) \geqslant 5^{1/2}$ for a Hilbert space H.

In extending the above, Kirk (1971) showed that a mapping $T: K \to K$ for which $T^n = I$ has a fixed point if $\|T^i x - T^i y\| \leqslant k \|x - y\|$, $x \, y \in K$, where k satisfies

$$(n-1)(n-2)k^2 + 2(n-1)k < n^2.$$

If T is k-lipschitzian with $k > 1$, then $\|T^i x - T^i y\| \leqslant k^{n-1} \|x - y\|$ for $i = 1, 2, \ldots, n-1$. Thus a k-lipschitzian mapping satisfying $T^n = I$ has fixed points if

$$(n-1)(n-2)k^{2(n-1)} + 2(n-1)k^{n-1} < n^2.$$

In terms of γ, the above translates to:

$$\gamma(0, 2, X) \geqslant 2;$$

$$[\gamma(0, n, X)]^{n-1} \geqslant (n-2)^{-1}[-1 + (n(n-1) - (n-1)^{-1})^{1/2}], \qquad n > 2.$$

We now give a simple example of an estimate for $\gamma(a, 2, X)$. Suppose $T \in \Phi(a, 2, k, K)$, and for $\alpha \in (0,1)$ and $x \in K$, set

$$z = (1 - \alpha)x + \alpha Tx;$$
$$u = (1 - \alpha)T^2 x + \alpha Tx. \qquad (17.16)$$

Then

$$\|z - Tz\| \leqslant \|z - u\| + \|u - Tz\|$$
$$\leqslant (1 - \alpha)\|x - T^2 x\| + (1 - \alpha)\|T^2 x - Tz\| + \alpha\|Tx - Tz\|$$
$$\leqslant (1 - \alpha)a\|x - Tx\| + (1 - \alpha)k\|Tx - z\| + \alpha k\|x - z\|$$
$$\leqslant [(1 - \alpha)a + (1 - \alpha)^2 k + \alpha^2 k]\|x - Tx\|$$
$$= h(\alpha)\|x - Tx\|.$$

The function h attains its infimum for $\alpha_0 = (a + 2k)/(4k)$, and

$$h(\alpha_0) = \frac{4k^2 + 4ak - a^2}{8k}.$$

Thus if

$$k < \tfrac{1}{2}[2 - a - (a^2 + (2-a)^2)^{1/2}], \tag{17.17}$$

then $h(\alpha_0) < 1$. Setting $F = (1 - \alpha_0)I + \alpha_0 T$, we have

$$\|F^2 x - Fx\| \leqslant h(\alpha_0)\|x - Fx\|$$

which implies that the sequence $\{F^n x\}$ of iterates of F at $x \in K$ converges to a point of Fix $F =$ Fix T provided (17.17) holds.

Another estimate for $\gamma(a, 2, X)$ may be obtained as follows. Beginning with ((17.16),

$$
\begin{aligned}
\|Tu - u\| &= \|\alpha Tu + (1-\alpha)Tu - \alpha Tx - (1-\alpha)T^2 x\| \\
&\leqslant \alpha k\|u - x\| + (1-\alpha)k\|u - Tx\| \\
&\leqslant \alpha k\|\alpha Tx + (1-\alpha)T^2 x - \alpha x - (1-\alpha)x\| + (1-\alpha)^2 k^2\|x - Tx\| \\
&\leqslant \alpha^2 k\|x - Tx\| + \alpha(1-\alpha)k\|x - T^2 x\| + (1-\alpha)^2 k^2\|x - Tx\| \\
&\leqslant [\alpha^2 k + \alpha(1-\alpha)ka + (1-\alpha)^2 k^2]\|x - Tx\| \\
&= g(\alpha)\|x - Tx\|.
\end{aligned}
$$

Simplifying, $g(\alpha) = \alpha^2 k(1 + k - a) + \alpha k(a - 2k) + k^2$, and g attains its infimum for $\alpha_1 = (2k - a)/2(1 + k - a)$, with

$$g(\alpha_1) = \frac{k(4k - a^2)}{4(1 + k - a)}.$$

It follows as above that T has a fixed point if $g(\alpha_1) < 1$. Routine calculations and the previous estimates yield

$$
\begin{aligned}
\gamma(a, 2, X) \geqslant \max\{ &\tfrac{1}{2}[2 - a + (a^2 + (2-a)^2)^{1/2}], \\
&\tfrac{1}{8}[a^2 + 4 + ((a^2 + 4)^2 - 64(a-1))^{1/2}]\}
\end{aligned} \tag{17.18}
$$

Curiously, it is possible to show that the first term provides a better estimate if $a \leqslant 2(2^{1/2} - 1)$, while the second is better for $a \in [2(2^{1/2} - 1), 2]$.

It is not known whether the estimate (17.18) is sharp; indeed, it is not even known whether $\gamma(0, 2, X) < +\infty$ or, equivalently, whether there exists a $(0, 2)$-rotative lipschitzian mapping (involution) on a bounded, closed, convex subset K of X which is fixed point free. However, continuous involutions of this type do exist.

Example 17.1 V. Klee (1953) has proved that any Banach space X is homeomorphic with the 'punctured' space $X \backslash \{0\}$. (Also see Bessaga and Pelczynski, 1957.) Let $h: X \to X \backslash \{0\}$ be such a homeomorphism and assume (as one may) that $hx = x$ for $x \in X$, $\|x\| \geqslant 1$. Now define $T: X \to X$ by taking

$$Tx = h^{-1}(-hx).$$

Then $T^2 = I$ on $B(0; 1)$ and Fix $T = \emptyset$ (since $Tx = x$ implies $hx = -hx = 0$). Since h transforms a complete space into an incomplete one, h^{-1} is not uniformly continuous (and probably neither is T).

We now give one of the very few examples known which provides an upper bound for $\gamma(a, 2, X)$.

Example 17.2 Let $X = \mathscr{C}[0, 1]$ and consider the following set K (which has been used in several previous examples):

$$K = \{x \in \mathscr{C}[0, 1]: 0 = x(0) \leqslant x(t) \leqslant x(1) = 1, t \in [0, 1]\}.$$

Let $e(t) \equiv t$ on $[0, 1]$, select $\alpha \in K$, $\alpha \neq e$, and define $T_\alpha: K \to K$ by setting

$$(T_\alpha x)(t) = \alpha(x(t)) = \alpha \circ x(t).$$

The mapping T_α inherits the behavior of α, e.g. if $k(\alpha) \leqslant k$ then $k(T_\alpha) \leqslant k$. Also, since any function $x \in K$ takes on all values between 0 and 1, T_α has a surprising property:

$$\begin{aligned}
\|T_\alpha x - x\| &= \max\{|\alpha(x(t)) - x(t)|: t \in [0, 1]\} \\
&= \max\{|\alpha(t) - t|: t \in [0, 1]\} \\
&= \|\alpha - e\| \\
&= d_1 > 0.
\end{aligned}$$

Thus any point $x \in K$ is displaced the same fixed distance d_1 by T_α. Also observe that for the iterates we have $T_\alpha^n = T_{\alpha^n}$ where $\alpha^n = \alpha \circ \cdots \circ \alpha$ n-times. Thus for each $x \in K$,

$$\|T_\alpha^n x - x\| = \|\alpha^n - e\| \equiv d_n > 0.$$

Therefore, since $1/d_1 \geqslant d_n/d_1$, any such mapping T_α is n-rotative for $n > 1/d_1$. However, since T_α is fixed point free, it must be the case that

$$k(T_\alpha) \geqslant \gamma(d_n/d_1, n, \mathscr{C}[0, 1]),$$

and this provides a rough estimate for γ from above.

Now consider the special case where

$$\alpha(t) = k \max\{0, t - 1 + k^{-1}\}.$$

Thus $d_1 = \|\alpha - e\| = 1 - k^{-1}$, $d_2 = \|\alpha^2 - e\| = 1 - k^{-2}$ and, in general, $d_n = 1 - k^{-n}$. Obviously, $k(T_\alpha) = k$ and, in general, $k(T_\alpha^n) = k^n$. Therefore for all

$x \in K$,

$$\|x - T_\alpha^2 x\| = 1 - k^{-2} = (1 + k^{-1})\|x - T_\alpha x\|$$

which shows that T_α is $(1 + k^{-1}, 2)$-rotative. Setting $a = 1 + k^{-1}$ we conclude:

$$\gamma(a, 2, \mathscr{C}[0, 1]) \leqslant (a - 1)^{-1}.$$

(Here, of course, $a > 1$. Aside from the observations regarding $\gamma(0, 2, X)$, no examples or estimates for γ are known in the case $a \leqslant 1$.)

Another special case illustrates a different type of singular behavior. Consider α given by

$$\alpha(t) = \begin{cases} 0 & \text{if } 0 \leqslant t \leqslant \epsilon \\ t - \epsilon & \text{if } \epsilon \leqslant t \leqslant 1 - \epsilon \\ 2t - 1 & \text{if } 1 - \epsilon \leqslant t \leqslant 1. \end{cases}$$

Here $|\alpha(t) - \alpha(s)| \leqslant 2|t - s|$ for $t, s \in [0, 1]$; thus $k(T_\alpha) = 2$ and $d_1 = \|\alpha - e\| = \epsilon$. However in addition, $d_2 = \|\alpha^2 - e\| = 2\epsilon$, $d_3 = \|\alpha^3 - e\| = 3\epsilon, \ldots, d_n = n\epsilon$ as long as $n\epsilon \leqslant 1 - \epsilon$. Thus T_α is highly non n-rotative for $n \leqslant (1 - \epsilon)\epsilon^{-1}$, yet T_α becomes n-rotative for larger n. However this rotativeness is too weak to yield a fixed point. (We note that actually, for $n \leqslant (1 - \epsilon)\epsilon^{-1}$ any orbital segment $\{T_\alpha^m x, T_\alpha^{m+1} x, \ldots, T_\alpha^{m+n} x\}$ lies on a metric line.)

One might expect better estimates for $\gamma(a, 2, X)$ when X exhibits nicer geometric structure and, indeed, some are known. (Cf., Goebel and Koter 1981b; Koter, 1986.) We remark in particular that T. Komorowski (1987) has shown that for a Hilbert space H,

$$\gamma(a, 2, H) \geqslant \frac{5^{1/2}}{(a^2 + 1)^{1/2}} \tag{17.19}$$

with

$$\liminf_{a \to 2^-} \frac{\gamma(a, 2, H) - 1}{2 - a} \geqslant 2/5.$$

The estimates were obtained by combining technical calculations similar to ones used above in the general case with properties of the Hilbert space norm. While (17.19) is consistent with our previous estimate $\gamma(0, 2, H) \geqslant 5^{1/2}$, our next observation indicates that this estimate probably is not sharp.

Remark $\gamma(0, 2, H) \geqslant (\pi^2 - 3)^{1/2} > 5^{1/2}$.

Proof This requires some additional facts about the geometry of H. It is known (Schaffer, 1976) that any rectifiable curve lying on the unit sphere S of H which joins two antipodal points has length at least π. Since lipschitzian curves are rectifiable and the radial projection on the unit ball of H is nonexpansive, we have the following: If $\gamma: [a, b] \to H$ is a rectifiable curve with $\gamma(a) = x$, $\gamma(b) = -x$, and $\|\gamma(t)\| \geq r$ for $t \in [a, b]$, then the length of γ is at least πr.

Now suppose K is a bounded, closed and convex subset of H with $T: K \to K$ k-lipschitzian and satisfying $T^2 = I$. Fix $\epsilon > 0$, and for any $x_0 \in K$ consider the curve $\gamma_{x_0}: [0, 1] \to H$ given by

$$\gamma_{x_0}(t) = T((1-t)x_0 + tTx_0) - (1-t)x_0 - tTx_0.$$

Thus $\gamma_{x_0}(0) = Tx_0 - x_0$ and $\gamma_{x_0}(1) = T^2x_0 - Tx_0 = x_0 - Tx_0 = -\gamma_{x_0}(0)$. If for some $t_0 \in [0, 1]$, $\|\gamma_{x_0}(t_0)\| < (1-\epsilon)\|Tx_0 - x_0\|$, set $x_1 = (1-t_0)x_0 + t_0 Tx_0$ and consider the curve γ_{x_1} which is obtained by replacing x_0 with x_1 in the definition of γ_{x_0}. Continue this procedure until either a sequence $\{x_n\}$ is obtained which converges to a fixed point of T, or until a point $x \in K$ is reached for which

$$\|\gamma_x(t)\| \geq (1-\epsilon)\|x - Tx\| \qquad \text{for all } t \in [0, 1].$$

Suppose the latter circumstance arises. Then the length of γ_x is at least $(1-\epsilon)d\pi$ where $d = \|x - Tx\|$. Take $\mu > 0$ and n sufficiently large to obtain

$$(1-\epsilon)d\pi - \mu \leq \sum_{i=1}^{n} \left\| \gamma_x\left(\frac{i}{n}\right) - \gamma_x\left(\frac{i-1}{n}\right) \right\|. \tag{17.20}$$

Setting $y_i = (1 - (i/n))x + (i/n)Tx$, we have

$$(1-\epsilon)d\pi - \mu \leq \sum_{i=1}^{n} \|(Ty_i - y_i) - (Ty_{i-1} - y_{i-1})\|$$

$$= \sum_{i=1}^{n} \|(Ty_i - Ty_{i-1}) - (y_i - y_{i-1})\|$$

$$\leq n^{1/2} \left(\sum_{i=1}^{n} \|(Ty_i - Ty_{i-1}) - (y_i - y_{i-1})\|^2 \right)^{1/2}$$

$$= n^{1/2} \left(\sum_{i=1}^{n} (\|Ty_i - Ty_{i-1}\|^2 - 2\langle Ty_i - Ty_{i-1}, y_i - y_{i-1}\rangle + (d/n)^2 \right)^{1/2}$$

$$\leq n^{1/2} \left(\sum_{i=1}^{n} (k^2 + 1)(d/n)^2 - (2/n) \sum_{i=1}^{n} \langle Ty_i - Ty_{i-1}, Tx - x\rangle \right)^{1/2}$$

$$= n^{1/2} ((k^2 + 1)(d^2/n) + 2(d^2/n))^{1/2}$$

$$= [(k^2 + 3)^{1/2}]d.$$

Since μ and ϵ are arbitrary, the iteration $\{x_1, x_2, \ldots\}$ can terminate only if $k \geq (\pi^2 - 3)^{1/2}$, and this completes the proof.

We conclude with the observation that many aspects of the theory described in this chapter seem preliminary in nature; the estimates obtained may only crudely approximate the actual situation.

18

The theorems of Brouwer and Schauder

We have been using the term 'fixed point property' (f.p.p.) as it relates to the class of nonexpansive mappings but, of course, this property may be applied to any class of mappings. The 'topological' fixed point property is of fundamental importance in the broader context of fixed point theory.

Definition 18.1 A topological space X is said to have the (*topological*) *fixed point property* (t.f.p.p. or f.p.p. if the meaning is clear) if each continuous mapping $T: X \to X$ has a fixed point.

It is not surprising that the above property is a topological invariant.

Lemma 18.1 *If X and Y are homeomorphic and if X has the* t.f.p.p., *then Y also has the* t.f.p.p.

Proof Let $h: X \to Y$ be a homeomorphism with $h(X) = Y$, and suppose $f: Y \to Y$ is continuous. Then $g = h^{-1} \circ f \circ h: X \to X$ and g is continuous; hence there exists $x \in X$ such that $gx = x$, implying $y = hx = fy$.

Another very useful observation about the t.f.p.p. is the following:

Lemma 18.2 *If a topological space X has the* t.f.p.p. *and if Y is a retract of X, then Y has the* t.f.p.p.

Proof. Let r be a retraction of X onto Y and let $f: Y \to Y$ be continuous. Then f may be extended to a continuous mapping $g = f \circ r: X \to X$. Any fixed point of g must also be a fixed point of f.

Probably the simplest observation about the t.f.p.p. is the fact that any continuous function $f: [-1, 1] \to [-1, 1]$ has a fixed point (i.e., $[-1, 1]$ has the t.f.p.p.). This is because the function $\varphi: x \mapsto x - fx$ satisfies $\varphi(-1) \leqslant 0 \leqslant \varphi(1)$ and therefore must assume the value 0. In view of Lemma 18.1, all compact intervals and simple arcs have the t.f.p.p. However, proving that even such simple sets as triangles in $\mathbb{R}^2$ also have the t.f.p.p. requires a much more sophisticated argument. This fact is a very

187

special case of the following celebrated theorem which was first proved in 1912 by L. E. J. Brouwer.

Theorem 18.1 (Brouwer) *The closed unit ball B^n in $\mathbb{R}^n$ has the f.p.p.*

Brouwer's Theorem is very useful in applications in analysis and its discovery has had a tremendous influence in the development of several branches of mathematics, most notably algebraic topology.

The simple formulation of Brouwer's Theorem belies the fact that it seems to require a 'nonelementary' (nonmetric) proof. All known proofs require facts not commonly used in metric fixed point theory. On the other hand, attempts to find more general versions of Brouwer's Theorem have given rise to several interesting questions of metric type which will be discussed in the next chapter. In view of this, we include a proof of Brouwer's Theorem here, and we have chosen the one which seems to us closest in flavor to the metric approach. This is the proof published in 1978 by J. Milnor (which yields, at the same time, another well-known theorem – the so-called 'Hairy Ball Theorem').

Throughout the remainder of the chapter B^n and S^{n-1} will denote, respectively, the unit ball and unit sphere in $\mathbb{R}^n$.

$$B^n = \{x \in \mathbb{R}^n : \|x\| \leqslant 1\};$$
$$S^{n-1} = \{x \in \mathbb{R}^n : \|x\| = 1\},$$

and we shall assume $\mathbb{R}^n$ is endowed with its standard inner product, $\langle x, y \rangle = \Sigma_{i=1}^n x_i y_i$ and norm, $\|x\| = \langle x, x \rangle^{1/2}$.

Let $A \subset \mathbb{R}^n$. A continuous mapping $f \colon A \to \mathbb{R}^n$ is said to be *of class C^1* if it has a continuous extension to an open neighborhood of A on which it is continuously differentiable.

To facilitate intuition, we shall often refer to the 'vector field' $f \colon A \to \mathbb{R}^n$. In doing so we think not of a mapping f which assigns to a point $x \in A$ a point $fx \in \mathbb{R}^n$, but rather a mapping which assigns to x the *vector* with initial point x and terminal point $x + fx$.

A mapping (vector field) $f \colon A \to \mathbb{R}^n$ is said to be *nonvanishing* if $fx \neq 0$ for all $x \in A$, and *normed* if $\|fx\| = 1$ for all $x \in A$ (i.e., if $f \colon A \to S^{n-1}$). The field $f \colon S^{n-1} \to \mathbb{R}^n$ is said to be *tangent* to S^{n-1} if $\langle x, fx \rangle = 0$ for $x \in S^{n-1}$.

We shall need the following fact.

Lemma 18.3 *If A is a compact subset of $\mathbb{R}^n$ and if $f \colon A \to \mathbb{R}^n$ is of class C^1 on A, then f satisfies, for some $L \geqslant 0$, the Lipschitz condition:*

$$\|fx - fy\| \leqslant L\|x - y\|, \qquad x, y \in A.$$

Proof For each $x \in A$ let U_x be an open ball of $\mathbb{R}^n$ which contains x and on the closure of which f is continuously differentiable. It is possible to cover A with a finite number, $U_1, \ldots, U_p$, of such balls. Set

$$c_{ij}^k = \max\left\{\left|\frac{\partial f_i x}{\partial x_j}\right| : x \in \bar{U}_k\right\}, \quad i, j = 1, \ldots, n; k = 1, \ldots, p.$$

Since each set U_k is convex, the Mean Value Theorem implies that for each $x, y \in U_k$,

$$\| fx - fy \| \leqslant \sum_{i=1}^{n} |f_i x - f_i y| \leqslant \sum_{i=1}^{n} \sum_{j=1}^{n} c_{ij}^k \| x - y \|$$

$$= \sum_{i,j=1}^{n} c_{ij}^k \| x - y \|.$$

On the other hand, a simple compactness argument shows that there exists $\epsilon > 0$ such that if $x, y \in A$ do not both lie in one of the sets $U_1, \ldots, U_p$, then $\| x - y \| \geqslant \epsilon$. Thus for such x, y,

$$\| fx - fy \| \leqslant \epsilon^{-1} \operatorname{diam} f(A) \| x - y \|.$$

The proof is completed by selecting

$$L = \max\{L_1, \ldots, L_p, \epsilon^{-1} \operatorname{diam} f(A)\},$$

where $L_k = \Sigma_{i,j=1}^{n} c_{ij}^k, k = 1, \ldots, p$.

Now assume $F: A \to \mathbb{R}^n$ is a vector field and for $t \in \mathbb{R}$ define $f_t : A \to \mathbb{R}^n$ by

$$f_t x = x + tFx, \qquad x \in A.$$

Lemma 18.4 *If A is a closed, bounded domain (i.e. the closure of a connected open set) in $\mathbb{R}^n$ and if $F: A \to \mathbb{R}^n$ is of class C^1 on A, then there exists an interval $\langle -\epsilon, \epsilon \rangle$ for some $\epsilon > 0$ on which the function*

$$\varphi : t \mapsto \operatorname{Vol} f_t(A)$$

is a polynomial of degree at most n.

(Vol in the above refers to the n-dimensional (Lebesgue measure) volume in $\mathbb{R}^n$.)

Proof Let L be the Lipschitz constant for F of Lemma 18.3 and take $\epsilon \in (0, L^{-1})$ so small that the determinant

$$\det\left[I + t\left(\frac{\partial F_i x}{\partial x_j}\right)\right] \tag{18.1}$$

remains positive for $x \in A$ and $|t| < \epsilon$. For such t, the Inverse Mapping Theorem implies that each f_t is a homeomorphism of class C^1, and thus Vol $f_t(A)$ may be evaluated by a well-known change of variables formula. The determinant (18.1) (the Jacobian of f_t) may be represented as

$$1 + ta_1(x) + t^2 a_2(x) + \cdots + t^n a_n(x)$$

where each a_i, $i = 1, \ldots, n$, is a continuous function. Thus

$$\text{Vol } f_t(A) = \alpha_0 + t\alpha_1 + t^2 \alpha_2 + \cdots + t^n \alpha_n$$

where

$$\alpha_i = \int_A a_i(x) \, dx$$

and

$$\alpha_0 = \text{Vol } A.$$

The next lemma deals with vector fields tangent to S^{n-1}.

Lemma 18.5 *Suppose* $F: S^{n-1} \to \mathbb{R}^n$ *is a normed vector field of class* C^1 *which is tangent to* S^{n-1}. *Then for* $t > 0$ *sufficiently small* $f_t: x \mapsto x + tFx$ *maps* S^{n-1} *onto the sphere centered at the origin of radius* $(1 + t^2)^{1/2}$, *i.e. onto* $(1 + t^2)^{1/2} S^{n-1}$.

Proof First extend the field F by setting, for $x \in \mathbb{R}^n \backslash \{0\}$,

$$\tilde{F}x = \|x\| F(x / \|x\|).$$

Let $A = \{x \in \mathbb{R}^n : \frac{1}{2} \leqslant \|x\| \leqslant \frac{3}{2}\}$ and assume $|t| < \min\{\frac{1}{3}, L^{-1}\}$, where L is the Lipschitz constant of $\tilde{F}$ on A. For fixed $z \in S^{n-1}$ define $G: A \to A$ by setting $Gx = z - t\tilde{F}x$. Then G is a contraction mapping, so by the Banach Contraction Principle (Theorem 2.1) G has a fixed point $x \in A$; thus $x + t\tilde{F}x = z$. Note that $\|x\| = (1 + t^2)^{-1/2}$, so $y = (1 + t^2)^{1/2} x \in S^{n-1}$. Thus $y + tFy = (1 + t^2)^{1/2} z$ and $f_t(S^{n-1}) = (1 + t)^{1/2} S^{n-1}$.

The Hairy Ball Theorem

The previous lemma dealt with vector fields tangent to S^{n-1}. If n is even, say $n = 2k$, then it is not difficult to construct concrete examples of such fields. One of the simplest is given by

$$F: (x_1, x_2, \ldots, x_{2k-1}, x_{2k}) \mapsto (x_2, -x_1, x_4, -x_3, \ldots, x_{2k}, -x_{2k-1}).$$

For n odd, the situation is completely different.

Theorem 18.2 (*The Hairy Ball Theorem – Weak Version*). *There are no normed vector fields of class C^1 tangent to S^{2k}.*

Proof Suppose such a vector field F does exist, let $0 < a < 1 < b$, and extend F, as in Lemma 18.5 to $\tilde{F}$ defined on the domain

$$A = \{x \in \mathbb{R}^n : a \leqslant \|x\| \leqslant b\}.$$

Note that $\tilde{F}$ is tangent to any sphere concentric with S^{2k} which is contained in A. As before, set $f_t x = x + t\tilde{F}x$. By Lemma 18.5, $f_t(A) = (1 + t^2)^{1/2} A$ for $t > 0$ sufficiently small. Thus

$$\text{Vol } f_t(A) = (1 + t^2)^{(2k+1)/2} \text{Vol } A.$$

However, $(1 + t^2)^{(2k+1)/2}$ does not coincide with any polynomial in a neighborhood of zero, contradicting the conclusion of Lemma 18.4.

We now prove a strong vesion of the above

Theorem 18.3 (*The Hairy Ball Theorem*) *There are no nonvanishing continuous vector fields tangent to S^{2k}.*

Proof Suppose there exists such a field F, and set $m = \min\{\|Fx\| : x \in S^{2k}\}$. Obviously $m > 0$. By the Weierstrass Approximation Theorem (recall Chapter 2), applied to each component of F, there exists a vector field $P : S^{2k} \to \mathbb{R}^{2k+1}$ for which

$$\|Px - Fx\| < m/2$$

for all $x \in S^{2k}$, where each component of P is a polynomial. Thus the field P is of class C^∞ and, moreover, P is nonvanishing since

$$\|Px\| \geqslant \|Fx\| - \|Px - Fx\| > m/2.$$

Now modify P by introducing the vector field Q defined by

$$Qx = Px - \langle Px, x \rangle x.$$

Then Q is also of class C^∞ and tangent to S^{2k}. Moreover,

$$\begin{aligned}
\|Qx\| &\geqslant \|Px\| - \|Qx - Px\| \\
&> (m/2) - |\langle Px, x \rangle| \\
&= (m/2) - |\langle Px - Fx, x \rangle| \\
&\geqslant (m/2) - \|Px - Fx\| \\
&> 0.
\end{aligned}$$

Normalization of Q (replacing Qx with $Qx/\|Qx\|$) now leads to a contradiction of Theorem 18.2.

We now turn to the proof of Brouwer's Theorem, and begin with some notation. The space $\mathbb{R}^n$ may be viewed as a subspace of $\mathbb{R}^{n+1}$ by identifying each point $x=(x_1,\ldots,x_n)\in\mathbb{R}^n$ with the point $(x_1,\ldots,x_n,0)\in\mathbb{R}^{n+1}$. Thus any point of $\mathbb{R}^{n+1}$ may be represented as (x,x_{n+1}) with $x\in\mathbb{R}^n$, $x_{n+1}\in\mathbb{R}$. The unit sphere $S^n\subset\mathbb{R}^{n+1}$ may be divided into two hemispheres – the upper (northern):

$$S^n_+ = \{(x,x_{n+1})\in S^n : x_{n+1}\geqslant 0\}$$

and lower (southern):

$$S^n_- = \{(x,x_{n+1})\in S^n : x^{n+1}\leqslant 0\}.$$

The unit sphere $S^{n-1}=S^n_+\cap S^n_-$ forms the 'equator', and the points $e_{n+1}=(0,\ldots,0,1)$ and $-e_{n+1}=(0,\ldots,0,-1)$ correspond, respectively, to the 'north' and 'south' poles.

Stereographic projection from e_{n+1} to S^n is the mapping $S_+:\mathbb{R}^n\to S^n$ which assigns to any $x\in\mathbb{R}^n$ the point of intersection of S^n with the open half-line starting at e_{n+1} and containing x. Specifically,

$$S_+x=(2x/(1+\|x\|^2),(\|x\|^2-1)/(\|x\|^2+1)).$$

This mapping is of class C^∞ and it transforms B^n onto S^n_-. Also, for $x\in S^{n-1}$, $S_+x=x$. In an analogous manner, the stereographic projection S_- from $-e_{n+1}$ to S^n is defined for $x\in\mathbb{R}^n$ by

$$S_-x=(2x/(1+\|x\|^2),(1-\|x\|^2)/(1+\|x\|^2)).$$

We are now in a position to prove the main result of this chapter.

Proof of Brouwer's Theorem We begin with the case $n=2k$. Suppose there exists a continuous mapping $f:B^{2k}\to B^{2k}$ which has no fixed point. Then the vector field F_1 defined by $F_1z=x-fx$ is nonvanishing on B^n, and at any point $x\in S^{2k-1}$ this field is directed 'outward', i.e.,

$$\langle F_1x,x\rangle=1-\langle x,fx\rangle>0.$$

Now let

$$Fx=x-[(1-\|x\|^2)/(1-\langle x,fx\rangle)]fx.$$

The vector field F is also nonvanishing on B^n. Indeed, if $Fx=0$ then the three points $0,x,fx$ are collinear which, in turn, implies $\langle x,fx\rangle x=\|x\|^2 fx$

and, since $Fx=0$, it follows that $fx=x$ which is a contradiction. Also, for $x \in S^{2k-1}$, $Fx=x$.

Now, for any $x \in B^n$ consider the segment which joins x with $x+Fx$, i.e., the set $\{x+tFx: t \in [0,1]\}$. The image of this set under the prescribed stereographic projection S_+ is a differentiable arc with initial point lying on the lower hemisphere S^{2k}_-:

$$t \mapsto S_+(x+tFx).$$

Now assign to each point $y=S_+x \in S^{2k}_-$ the vector tangent to this arc at y. More precisely, define the vector field T_- on S^{2k}_- by

$$T_-y = \frac{d}{dt}S_+(x+tFx)\bigg|_{t=0}$$

$$= \frac{2}{(1+\|x\|^2)^2} \langle (1+\|x\|^2)Fx - 2\langle x, Fx \rangle x, 2\langle x, Fx \rangle \rangle.$$

Then T_- is nonvanishing, continuous, and tangent to S^{2k}_-. On the equator T_- is directed upward, i.e., $T_-y=e_{n+1}$ for $y \in S^{2k-1}$.

In a manner entirely analogous to the above, take the vector field $-F$ instead of F and project corresponding arcs from $-e_{n+1}$ by S_- to obtain the vector field T_+ defined on S^{2k}_+ by

$$T_+y = \frac{d}{dt}S_-(x-tFx)\bigg|_{t=0}$$

$$= \frac{2}{(1+\|x\|^2)^2} \langle 2\langle x, Fx \rangle - (1+\|x\|^2)Fx, 2\langle x, Fx \rangle \rangle.$$

This vector field is also directed upward on the equator and $T_+y=T_-y$ for $y \in S^{2k-1}$.

Now, for $y \in S^{2k}$, set

$$Ty = \begin{cases} T_-y & \text{for } y \in S^{2k}_- \\ T_+y & \text{for } y \in S^{2k}_+. \end{cases}$$

Then T is a continuous, nonvanishing, vector field tangent to S^{2k}, which is a contradiction. Therefore f must have a fixed point, so the proof of the theorem in the case $B^n=B^{2k}$ is complete.

To prove the result for $B^n=B^{2k-1}$, it suffices to observe that if the continuous mapping $f: B^{2k-1} \to B^{2k-1}$ is fixed point free then so also is the mapping $g: B^{2k} \to B^{2k}$ defined by

$$g(x, x_{2k}) = (fx, 0).$$

This contradicts the first part of the proof.

Some simple consequences of Brouwer's Theorem

There are several interesting facts which are either consequences of Brouwer's Theorem or equivalent to it.

Theorem 18.4 *The unit sphere S^{n-1} is not a retract of B^n.*

Proof If $R: B^n \to S^{n-1}$ were a retraction then the mapping

$$-R: B^n \to S^{n-1} \subset B^n$$

would be fixed point free, contradicting Brouwer's Theorem.

The above theorem, which may also be derived from the simple observation that S^{n-1} does not have the f.p.p., is not only a consequence of Brouwer's Theorem, but actually equivalent to it. For, suppose Theorem 18.4 is known, and suppose $f: B^n \to B^n$ is a continuous mapping which is fixed point free. Then for each $x \in B^n$ consider the half-line through x with direction $u(x) = (x - fx)/\| x - fx \|$, and let Rx be the point at which this line intersects S^{n-1}. Then simple calculations show that

$$Rx = x + \lambda(x)u(x)$$

where

$$\lambda(x) = -\langle x, u(x) \rangle + (1 - \| x \|^2 + \langle x, u(x) \rangle^2)^{1/2};$$

thus R is a retraction of B^n onto S^{n-1}, a fact which contradicts Theorem 18.4.

The next result might appear even stronger than Brouwer's Theorem, but it also is merely equivalent to it.

Theorem 18.5 *If $F: B^n \to \mathbb{R}^n$ is a continuous vector field then either F vanishes at a point $y \in B^n$ or there exists a point $z \in S^{n-1}$ and a number $\lambda > 0$ such that $Fz = \lambda z$.*

Proof Define $f: B^n \to \mathbb{R}^n$ by

$$fx = \begin{cases} -F(4\| x \| x) & \text{for } \| x \| \leqslant \tfrac{1}{2}, \\ (2\| x \| - 1)x - (2 - 2\| x \|)F\left(\dfrac{x}{\| x \|} \right) & \text{for } \| x \| \geqslant \tfrac{1}{2}. \end{cases}$$

Obviously f is continuous and $fx = x$ for $x \in S^{n-1}$. Also, it must be the case that $fx = 0$ for some $x \in B^n$ for, otherwise, the mapping R given by

$Rx = fx/\|fx\|$ would be a retraction of B^n onto S^{n-1}. But if $fx = 0$, then either $\|x\| \leqslant \frac{1}{2}$ and $F(4\|x\|x) = 0$ or $\|x\| > \frac{1}{2}$ and for $z = (x/\|x\|) \in S^{n-1}$, $Fz = \lambda z$ with

$$0 < \lambda = \frac{2\|x\| - 1}{2(1 - \|x\|)} \|x\|.$$

To see that Brouwer's Theorem may be derived from the above, suppose $f: B^n \to B^n$ is fixed point free. Then, in view of the fact that $\langle x, Fx \rangle = \langle x, fx \rangle - 1 < 0$ for $x \in S^{n-1}$, the nonvanishing vector field given by $Fx = fx - x$ contradicts Theorem 18.5.

Another version of Theorem 18.5 is the following.

Theorem 18.6 *If $f: B^n \to \mathbb{R}^n$ is a continuous mapping with no fixed points in B^n, then there exists $z \in S^{n-1}$ and a number $\lambda > 1$ such that $fz = \lambda z$.*

Our next result requires the notion of contractibility.

Definition 18.2 A topological space X is said to be *contractible* (to a point $z \in X$) if there exists a continuous function (a homotopy) $H: X \times [0, 1] \to X$ such that $H(x, 0) = x$ and $H(x, 1) = z$ for all $x \in X$.

It is easy to see that if a topological space X is contractible to some point $z \in X$, then it is contractible to any point $y \in X$.

Theorem 18.7 *The sphere S^{n-1} is not contractible.*

Proof Suppose S^{n-1} is contractible to $z \in S^{n-1}$ via the function H. Then, for example, the mappings $R_1, R_2: B^n \to S^{n-1}$ given by

$$R_1 x = \begin{cases} z & \text{for } \|x\| \leqslant \frac{1}{2}, \\ H(x/\|x\|, 2(1 - \|x\|)) & \text{for } \|x\| \geqslant \frac{1}{2}; \end{cases}$$

$$R_2 x = \begin{cases} z & \text{for } x = 0, \\ H(x/\|x\|, 1 - \|x\|) & \text{for } x \neq 0 \end{cases}$$

would be retractions of B^n onto S^{n-1}, a fact which contradicts Theorem 18.4.

As before, the above theorem is actually equivalent to Brouwer's Theorem. To see this, it suffices to observe that if R were a retraction of B^n onto S^{n-1}, then the mapping H defined by

$$H(x, t) = R((1 - t)x)$$

would contract S^{n-1} to the point $R(0)$.

The above facts may be all summed up as follows:

Theorem 18.8 *The following statements are equivalent.*

(a) B^n *has the fixed point property ($f.p.p.$)*
(b) S^{n-1} *is not a retract of B^n.*
(c) S^{n-1} *is not contractible.*

Our next consequence of Brouwer's Theorem requires the following.

Lemma 18.6 *Each nonempty, closed and convex subset C of $\mathbb{R}^n$ is a retract of $\mathbb{R}^n$.*

Proof This may be demonstrated in several ways but the following is one of the simplest. For any $x \in \mathbb{R}^n$ there exists a unique point $y = Rx \in C$ such that

$$\|x - y\| = \inf\{\|x - u\| : u \in C\}.$$

As seen earlier (Chapter 12), the mapping R is nonexpansive, hence a retraction of $\mathbb{R}^n$ onto C.

The following is a consequence of the above.

Theorem 18.9 (*Brouwer's Theorem – Strong Version*). *Each nonempty, bounded, closed and convex subset C of $\mathbb{R}^n$ has the $f.p.p.$*

Proof Since C is bounded it is contained in some ball in $\mathbb{R}^n$ and hence is a retract of that ball. Thus the result follows from Lemma 18.2.

Finally, we remark that since any finite dimensional Banach space X is isomorphic to $\mathbb{R}^n$ with $n = \dim X$, the above has a slightly more general form.

Theorem 18.9′ *Any nonempty, bounded, closed and convex subset of a finite dimensional Banach space has the $f.p.p.$*

An important generalization of Brouwer's Theorem was discovered in 1930 by J. Schauder.

Theorem 18.10 *Any nonempty, compact, convex subset K of a Banach space has the t.f.p.p*

Proof Let $T: K \to K$ be a continuous mapping and take $\epsilon > 0$. Since K is compact there exists a finite ϵ-net $\{a_1, \ldots, a_p\} \subset K$. Define a collection $\{m_i\}$ of real-valued functions on K by taking

$$m_i(x) = \begin{cases} 0 & \text{if } \|x - a_i\| \geq \epsilon, \\ \epsilon - \|x - a_i\| & \text{if } \|x - a_i\| \leq \epsilon, \end{cases} \qquad i = 1, \ldots, p,$$

and define the function $\varphi: K \to K_0 = K \cap \text{span}\{a_1, \ldots, a_p\}$ by taking

$$\varphi(x) = \left(\sum_{i=1}^{p} m_i(x) a_i \right) \bigg/ \left(\sum_{i=1}^{p} m_i(x) \right).$$

Obviously φ is continuous, and for any $x \in K$, $\|\varphi(x) - x\| \leq \epsilon$.

Now consider the modified mapping $\tilde{T}: K \to K$ defined by $\tilde{T} = \varphi \circ T$. Thus $\tilde{T}: K_0 \to K_0$ and by Theorem 18.9' $\tilde{T}$ has a fixed point $x \in K_0$. From this we have

$$\|x - Tx\| \leq \|x - \tilde{T}x\| + \|\tilde{T}x - Tx\|$$
$$= \|\varphi \circ Tx - Tx\|$$
$$\leq \epsilon.$$

Therefore, $\inf\{\|x - Tx\| : x \in K\} = 0$, and since K is compact and T continuous, Fix $T \neq \emptyset$.

A modified version of the above is also useful.

Theorem 18.10' *Let K be a nonempty, closed, convex subset of a Banach space and suppose $T: K \to K$ is continuous with $\overline{T(K)}$ compact. Then T has a fixed point in K.*

Proof Since the set $K_0 = \overline{\text{conv}}\, T(K)$ is compact, convex and T-invariant, this result is an immediate consequence of Theorem 18.10.

Mappings T having relatively compact ranges $T(K)$ are usually called compact mappings. There is, in fact, an even larger class of mappings, one which bridges the gap between the topological and metric theory, for which all bounded and convex sets have the fixed point property. This class was introduced by Darbo (1955). For the sake of completeness we also include Darbo's result.

Theorem 18.11 *Let K be a nonempty, bounded, closed, convex subset of a Banach space X and let α denote the Kuratowski measure of noncompactness in X. Suppose $T: K \to K$ is a continuous mapping which satisfies for fixed $k \in [0, 1)$ and any $E \subset K$,*

$$\alpha(T(E)) \leqslant k\alpha(E).$$

Then T has a fixed point in K.

Proof Let $K_0 = K$ and define

$$K_{n+1} = \overline{\text{conv}}\, T(K_n), \qquad n = 0, 1, \ldots,$$

Then $\{K_n\}$ is a decreasing sequence of T-invariant sets and, moreover,

$$\begin{aligned}
\alpha(K_{n+1}) &= \alpha(\overline{\text{conv}}\, T(K_n)) \\
&= \alpha(T(K_n)) \\
&\leqslant k\alpha(K_n);
\end{aligned}$$

hence $\alpha(K_{n+1}) \leqslant k^{n+1}\alpha(K_0)$ which implies $\lim_{n \to \infty} \alpha(K_n) = 0$. Therefore the set

$$K_\infty = \bigcap_{n=1}^{\infty} K_n$$

is nonempty and compact. Since K_∞ is also closed, convex and T-invariant, Schauder's Theorem implies Fix $T \neq \emptyset$.

Obviously the Kuratowski measure of noncompactness α could be replaced with any such measure (e.g., the Hausdorff measure) having similar properties. We remark that mappings satisfying the assumptions of Darbo's Theorem include transformations of the form $T = T_1 + T_2$ where T_1 is a compact mapping and T_2 a contraction mapping.

Multivalued mappings Brower's Theorem has also been extended to the multivalued case. The basic results in this direction are due to S. Kakutani (1941), Bohnenblust and Karlin (1950), and Ky Fan (1960/61) for, respectively, the finite dimensional and compact cases. In order to discuss these extensions it is necessary to review facts about semicontinuity, which we do in a special setting sufficient to our purpose.

Let K be a compact, convex subset of a Banach space X and let $(\mathscr{K}, D)$ be the metric space consisting of all nonempty, bounded, closed subsets of K endowed with the Hausdorff metric D. A mapping $F: K \to \mathscr{K}$ is said to be *upper semicontinuous* if for any sequences $\{x_n\}$ and $\{y_n\}$ with $\{x_n\}$ in K,

$y_n \in Tx_n$, the conditions $\lim_{n\to\infty} x_n = x$ and $\lim_{n\to\infty} y_n = y$ imply $y \in Fx$. F is said to be *lower semicontinuous* if for any sequence $\{x_n\}$ in K with $\lim_{n\to\infty} x_n = x$ there exists a sequence $\{y_n\}$ with $y_n \in Fx_n$ such that $\lim_{n\to\infty} y_n = y \in Fx$. Finally, F is said to be continuous if it is both upper and lower semicontinuous (and it is easy to see that this formulation is equivalent to continuity with respect to the Hausdorff metric).

In general a continuous mapping $F: K \to \mathcal{K}$ does not always admit a *continuous selection*, i.e., a continuous mapping $f: K \to K$ for which $fx \in Fx$, $x \in K$ (cf., Example 18.1 below). However the situation is different for convex-valued mappings.

Lemma 18.7 *If $F: K \to \mathcal{K}$ is a continuous convex-valued mapping, then F has a continuous selection $f: K \to K$.*

Proof Since K is compact the smallest subspace of X containing K is separable so, for simplicity, we may assume X is separable. Since any separable space has an equivalent strictly convex norm (see Zizler's Theorem, Chapter 6) we may assume the norm of X is strictly convex at the outset and for $p \in X$ define $f_p: K \to K$ by taking $f_p x$ to be the unique point in Fx which is nearest p:

$$f_p x = \operatorname{Proj}_{Fx} p.$$

A routine argument shows that f_p is continuous.

(The above fact represents a mere starting point for the rich selection theory developed by Michael (1956a,b; 1957) (see also Aubin and Cellina, 1984). In particular it is known that in our specialized setting convex-valued lower semicontinuous mappings always have continuous selections.)

One consequence of Lemma 18.7 is that any continuous convex-valued mapping $F: K \to \mathcal{K}$ has a fixed point, since by Schauder's Theorem there exists $x \in K$ for which $x = fx \in Fx$.

Now supose $\{F_n\}$ is a sequence of continuous, convex-valued mappings of K into $\mathcal{K}$ such that for each $x \in K$,

$$F_1 x \supseteq F_2 x \supseteq F_3 x \supseteq \cdots,$$

and set $Fx = \bigcap_{n=1}^{\infty} F_n x$, $x \in K$. The mapping F, although not necessarily continuous, is always upper semicontinuous and also convex-valued. A straightforward compactness argument shows that F also always has a fixed point. Conversely, the following is true.

Lemma 18.8 *If $F: K \to \mathcal{K}$ is upper semicontinuous and convex-valued, then*

F can be represented as the intersection of a descending sequence of convex-valued continuous functions.

Proof Let $\epsilon>0$, and for $x\in K$ set $V(x,\epsilon)=B(x;\epsilon)\cap K$. Then for $y\in V(x,\epsilon)$, $\|y-x\|=\alpha\epsilon$ for some $\alpha\in[0,1]$. For such y, set

$$U(y)=\bigcup_{\beta\in[\alpha,1]}(\beta Fx+(1-\beta)Fy)$$

and

$$F_\epsilon x=\overline{\text{conv}}\bigcup_{y\in V(x,\epsilon)}U(y).$$

It is easy to see that F_ϵ is continuous with $\epsilon'<\epsilon$ implying $F_{\epsilon'}x\subseteq F_\epsilon x$, $x\in K$. Thus the sequence $\{F_{(1/n)}\}$ satisfies the requirements of the lemma.

The above facts yield the following

Theorem 18.12 *If K is a nonempty, compact, convex subset of a Banach space then any upper semicontinuous, convex-valued mapping of K into the nonempty, closed subsets of K has a fixed point.*

We conclude by showing that the assumption that the mapping be convex-valued is essential.

Example 18.1 Consider the plane $\mathbb{R}^2$ with its standard euclidean norm, and assign to each point $x\in\mathbb{R}^2$ and $\epsilon>0$ a circle $T_\epsilon x$ with center x and radius ϵ. Thus

$$T_\epsilon x=\{y\in\mathbb{R}^2:\|x-y\|=\epsilon\}.$$

Then T_ϵ is a compact-valued mapping which is an isometry relative to the Hausdorff metric:

$$D(T_\epsilon x,T_\epsilon y)=\|x-y\|,\qquad x,y\in\mathbb{R}^2.$$

Obviously, a continuous selection of T_ϵ may be obtained by assigning to each $x=(x_1,x_2)\in\mathbb{R}^2$ its ϵ-shift $(x_1+\epsilon,x_2)$.

Now let Δ be the closed unit disc:

$$\Delta=\{x=(x_1,x_2):x_1^2+x_2^2\leqslant1\},$$

and let $\epsilon\in(0,1)$. Then the multivalued mapping F defined on Δ by taking

$$Fx=T_\epsilon x\cap\Delta,\qquad x\in\Delta,$$

is neither an isometry nor nonexpansive; yet F is continuous. Since for any

$x \in \Delta$, dist$(x, Fx) = \epsilon$, F does not have a fixed point (and, in particular, F does not admit a continuous selection).

As a final observation, consider the unit square

$$\Delta_\infty = \{x = (x_1, x_2): |x_1| \leqslant 1, |x_2| \leqslant 1\},$$

and for $x \in \Delta_\infty$ set

$$\tilde{F}x = T_\epsilon x \cap \Delta_\infty.$$

Then $\tilde{F}$ is upper semicontinuous, yet not continuous.

Further information regarding the topics of this chapter may be found, for example, in Browder, 1976; Dunford and Schwartz, 1957; Smart, 1974; Dugundji and Granas, 1982; and Istratescu, 1981.

19

Lipschitzian mappings

As we have seen in previous examples, it is in general not possible, even in limited ways, to extend the Brouwer and Schauder theorems to non-compact settings. However, attempts to find such extensions have raised some intriguing 'geometrical' problems which remain open. We begin with some historical observations.

The famous collection of mathematical problems known as *The Scottish Book* (see Mauldin, 1981 for details) contains a question (Problem 36) raised around 1935 by S. Ulam which reads: 'Can one transform continuously the solid sphere of a Hilbert space into its boundary such that the transformation should be the identity on the boundary of the ball?' An addendum states: 'There exists a transformation with the required property given by Tychonoff.'

Ulam's problem is probably one of the simplest appearing in *The Scottish Book*. In our terminology it reads: Is the unit sphere in a Hilbert space a retract of its unit ball? The answer to this question is most commonly attributed to S. Kakutani who, in 1943, presented several examples of continuous self-mappings of the unit ball in Hilbert space without fixed points, any of which may be used to provide an answer to Ulam's question.

Example 19.1 (Kakutani's Construction) Consider two models of Hilbert space: $l^2(\mathbb{N})$ and $l^2(\mathbb{Z})$. For the unit ball B in $l^2(\mathbb{N})$ the mapping $T: B \to B$ defined by

$$T(x_1, x_2, \ldots) = ((1 - \|x\|^2)^{1/2}, x_1, x_2, \ldots)$$

is clearly continuous with Fix $T = \emptyset$. The same is true for the mapping $T_\epsilon: B \to B$ defined for $\epsilon > 0$ by

$$T_\epsilon(x_1, x_2, \ldots) = (\epsilon(1 - \|x\|), x_1, x_2, \ldots).$$

Moreover, in the latter case $\|T_\epsilon x - T_\epsilon y\| \leq (1 + \epsilon^2)^{1/2} \|x - y\|$ for $x, y \in B$; hence $k(T_\epsilon) = (1 + \epsilon^2)^{1/2}$.

For $l^2(\mathbb{Z})$ the shift operator

$$S: x = \{x_i\}_{i \in \mathbb{Z}} \mapsto \{x_{i+1}\}_{i \in \mathbb{Z}}$$

is an isometry of B onto B. Thus the mapping $T: B \to B$ defined by

$$Tx = \tfrac{1}{2}(1 - \|x\|)e_0 + Sx,$$

where $e_0 = \{\delta_{i,0}\}$, is a lipschitzian *homeomorphism B onto B* with a lipschitzian inverse, yet Fix $T = \emptyset$.

Any of the above mappings may be used to produce a retraction R of the unit ball onto its boundary via the method described in the previous chapter: For $x \in B$ take

$$Rx = x + \lambda(x)u(x)$$

where

$$u(x) = \frac{x - Tx}{\|x - Tx\|}; \qquad \lambda(x) = -\langle x, u(x)\rangle + (1 - \|x\|^2 + \langle x, u(x)\rangle^2)^{1/2}.$$

Another nice solution to Ulam's question, and one which holds in an arbitrary Banach space X, is given by the result of Klee (mentioned in Example 17.1) that there always exists a homeomorphism $h: X \to X \setminus \{0\}$ with $hx = x$ for $\|x\| \geq 1$. The required retraction is then given by $Rx = hx/\|hx\|$, $x \in B$. C. Bessaga (1966) (see also Bessaga and Pelczynski, 1975) showed that this mapping h may even be assumed smooth of class C^∞; thus very regular retractions exist which provide an answer to Ulam's question.

Note that given a retraction R of the unit ball onto its boundary, the mapping $-R$ also maps the unit ball onto its boundary and is fixed point free. Thus the unit ball in any infinite dimensional Banach space fails to have the fixed point property (f.p.p.) for continuous mappings. In fact, an even stronger result is given in Klee (1955).

Theorem 19.1 *For any noncompact, closed, convex subset K of a Banach space there exists a continuous mapping $T: K \to K$ with Fix $T = \emptyset$.*

The remainder of this chapter is devoted to a proof of the following two generalizations of the results discussed above.

Theorem 19.2 *For any infinite dimensional Banach space X there exists a lipschitzian retraction of the unit ball onto the unit sphere.*

Theorem 19.3 *For any noncompact, bounded, closed and convex subset K of a Banach space there is a lipschitzian mapping $T: K \to K$ for which*

$$\inf\{\|x - Tx\| : x \in K\} = d > 0.$$

Theorem 19.2 was first proved by Nowak (1979) using a complicated construction that was subsequently somewhat simplified by Benyamini and Sternfeld (1983). Here we derive Theorem 19.2 from a proof of Theorem 19.3 due to Lin and Sternfeld (1985).

The proof requires several preliminary technical lemmata.

Lemma 19.1 *Let* (M, ρ) *be a metric space and let* Z *be a closed subset of* M. *Suppose* $f : Z \to \mathbb{R}$ *is a function satisfying the Lipschitz condition*

$$|fx - fy| \leqslant k\rho(x, y), \qquad x, y \in Z. \tag{19.1}$$

Then f *has an extension* $\tilde{f} : M \to \mathbb{R}$ $(\tilde{f}|_Z = f)$ *which satisfies the same Lipschitz condition. Moreover, if* $a \leqslant fx \leqslant b$ *for all* $x \in Z$, *then* $\tilde{f}$ *may be chosen so that* $a \leqslant \tilde{f}x \leqslant b$ *for all* $x \in M$.

Proof It suffices to observe that the function $\tilde{f}_1$ defined by

$$\tilde{f}_1 x = \sup\{fy - k\rho(x, y) : y \in Z\}, \qquad x \in M,$$

has the first property. The second property holds upon taking

$$\tilde{f}x = \max\{a, \min\{b, \tilde{f}_1 x\}\}, \qquad x \in M.$$

Lemma 19.2 *Let* (M, ρ) *be a metric space and let* Z *be a closed subset of* M. *If* $f : Z \to l^\infty(\mathbb{N}_0)$ *(where* $\mathbb{N}_0 = \{0, 1, 2, \ldots, \}$*) is a function satisfying*

$$\|fx - fy\| \leqslant k\rho(x, y), \qquad x, y \in Z, \tag{19.2}$$

then f *has an extension* $\tilde{f} : M \to l^\infty(\mathbb{N}_0)$ *which satisfies the same condition.*

Proof Since $l^\infty(\mathbb{N}_0)$ consists of bounded sequences with the supremum norm, the function f is represented by a sequence $\{f_n\}$ of 'coordinate' functions each of which satisfies (19.1). Thus Lemma 19.2 follows upon applying Lemma 19.1 to each f_n.

Lemma 19.3 *Let* $x_1, \ldots, x_n$ *be points of a Banach space* X *satisfying, for* $1 \leqslant i \leqslant n$, $\|x_i\| \leqslant b$, $\|x_1\| \geqslant a$ *and* $\mathrm{dist}(x_i, \mathrm{span}\{x_1, \ldots, x_{i-1}\}) \geqslant a > 0$. *Then there exists a constant* $A > 0$ *such that if* $y \in \mathrm{span}\{x_1, \ldots, x_n\}$ *with* $y = \sum_{i=1}^n \alpha_i x_i$, *then*

$$A^{-1}\max\{|\alpha_i| : 1 \leqslant i \leqslant n\} \leqslant \|y\| \leqslant A \max\{|\alpha_i| : 1 \leqslant i \leqslant n\}.$$

It is important to note that the constant A above is universal for all Banach spaces, depending only on the constants a and b and the dimension n.

Next we observe that any noncompact, bounded set contains an infinite sequence whose terms exhibit the behavior of the set $\{x_1, \ldots, x_n\}$ in Lemma 19.3.

Lemma 19.4 *Let K be a noncompact, bounded, closed subset of a Banach space X with $0 \in K$. Then there exists a number $r > 0$ and a sequence $\{x_n\}$ of points of K such that for all $n \in \mathbb{N}$, $\text{dist}(x_{n+1}, \text{span}\{x_1, \dots, x_n\}) \geq r > 0$.*

Proof Let $\chi(K) > 0$ be the Hausdorff measure of noncompactness of K and let $\epsilon > 0$ satisfy $\epsilon < \chi(K)$. We show that it suffices to take $r = \chi(K) - \epsilon$.

Obviously there exists a point $x_1 \in K$ with $\|x_1\| \geq r$. Suppose that points $x_i \in K$ have already been obtained satisfying $\text{dist}\{x_{i+1}, \text{span}\{x_1, \dots, x_i\}\} \geq r$, $1 \leq i \leq n$. We have already seen that $\chi(K) = \text{dist}(K, \mathcal{N})$ where $\mathcal{N}$ is the family of compact subsets of K and the distance is taken in the Hausdorff metric sense (see Chapters 2 and 8). Thus K must contain a point x_{n+1} such that $\text{dist}(x_{n+1}, \text{span}\{x_1, \dots, x_n\}) \geq r$. This completes the induction.

In view of the above, any bounded, closed, noncompact set K with $0 \in K$ contains a sequence $\{x_n\}$ satisfying, for each $n \in \mathbb{N}$ and $\epsilon > 0$ sufficiently small,

$$\|x_n\| \leq b = \text{diam}(K);$$
$$\text{dist}(x_{n+1}, \text{span}\{x_1, \dots, x_n\}) \geq \chi(K) - \epsilon > 0. \tag{19.3}$$

We now pass to a special construction in $l^\infty(\mathbb{N}_0)$. For each $n \in \mathbb{N}_0$ let e_n denote the standard unit vector $\{\delta_{in}\} = (0, \dots, 0, 1, 0, \dots)$ (where 1 is in the nth position). Define a sequence $\{\Delta_n\}$ of triangles as follows:

$$\Delta_n = \text{conv}\{0, e_n, e_{n+1}\}$$
$$= \{(0, \dots, 0, \xi_n, \xi_{n+1}, 0, \dots) : \xi_n \geq 0, \xi_{n+1} \geq 0, \xi_n + \xi_{n+1} \leq 1\},$$

and let

$$\Delta = \bigcup_{n=0}^{\infty} \Delta_n.$$

Now let X be a Banach space and let K be a noncompact, bounded, closed, convex subset of X with $0 \in K$. Suppose $\{x_n\}, n = 0, 1, \dots,$ is a sequence in K satisfying (19.3). As above let $\{\Delta_n(K)\}$ denote the sequence of triangles defined by

$$\Delta_n(K) = \text{conv}\{0, x_n, x_{n+1}\},$$

and define

$$\Delta(K) = \bigcup_{n=0}^{\infty} \Delta_n(K).$$

Then we have:

Lemma 19.5 *There exists a homeomorphism H which maps Δ onto $\Delta(K)$ for which both H and H^{-1} are lipschitzian.*

Proof Define H in the following natural way. For each $n \in \mathbb{N}$, set

$$H(\xi_n e_n + \xi_{n+1} e_{n+1}) = \xi_n x_n + \xi_{n+1} x_{n+1}.$$

The conclusion of Lemma 19.5 is now a direct consequence of Lemma 19.3.

Lemma 19.6 *There exists a retraction $R: l^\infty(\mathbb{N}_0) \to \Delta$ which is lipschitzian on $l^\infty(\mathbb{N}_0)$.*

Proof Let $e = (1, 1, \dots) \in l^\infty(\mathbb{N}_0)$. For $x \in l^\infty(\mathbb{N}_0)$, set

$$E(x) = \{\epsilon \geqslant 0: (x - \epsilon e) \vee 0 = \{\max\{x_i - \epsilon, 0\}\} \in \Delta\}.$$

Then $E(x) \neq \emptyset$ since $\|x\| \in E(x)$, so we may set

$$\epsilon(x) = \inf\{\epsilon \geqslant 0: \epsilon \in E(x)\}.$$

Observe that for $x, y \in l^\infty(\mathbb{N}_0)$,

$$x \leqslant y + \|x - y\| e$$

in the sense: $x_i \leqslant y_i + \|x - y\|$, $i \in \mathbb{N}_0$. Thus $\epsilon(y) + \|x - y\| \in E(x)$ and then $\epsilon(x) \leqslant \epsilon(y) + \|x - y\|$. By symmetry we obtain

$$|\epsilon(x) - \epsilon(y)| \leqslant \|x - y\|.$$

Therefore if $R: l^\infty(\mathbb{N}_0) \to \Delta$ is defined by $Rx = (x - \epsilon(x)e) \vee 0$ then R is a retraction for which

$$\|Rx - Ry\| \leqslant 2\|x - y\|, \qquad x, y \in l^\infty(\mathbb{N}_0).$$

As an easy consequence of the above we have:

Lemma 19.7 *If K is a closed, convex, noncompact subset of a Banach space X, with $0 \in K$, and $\Delta(K)$ defined as above, then there exists a lipschitzian retraction of K onto $\Delta(K)$.*

Proof Let H^{-1} be the inverse of the homeomorphism $H: \Delta \to \Delta(K)$ of Lemma 19.5. By Lemma 19.2 H^{-1} has a lipschitzian extension $\tilde{H}^{-1}: K \to l^\infty$. The desired retraction $R_K: K \to \Delta(K)$ is now given by $R_K = H \circ R \circ \tilde{H}^{-1}$, where R is the retraction of Lemma 19.6.

Proof of Theorem 19.3 In addition to the above lemmas, the proof is based on the following two observations. First, suppose $T_0: \Delta(K) \to \Delta(K)$ is a lipschitzian mapping with $\inf\{\|x - T_0 x\|: x \in \Delta(K)\} = d > 0$. Then the mapping

$T_1: K \to K$ defined by $T_1 = T_0 \circ R_K$ is also lipschitzian with

$$\inf\{\|x - Tx\| : x \in K\} \geqslant d_1$$

where $0 < d_1 \leqslant d$. Second, suppose $T_2: \Delta \to \Delta$ is a lipschitzian mapping with $\inf\{\|x - T_2 x\| : x \in \Delta\} > 0$. Then the mapping $T_0: \Delta(K) \to \Delta(K)$ where $T_0 = H \circ T_2 \circ H^{-1}$ is also lipschitzian with

$$\inf\{\|x - T_0 x\| : x \in \Delta(K)\} > 0.$$

Therefore, in order to complete the proof it suffices to construct a lipschitzian mapping $T_2: \Delta \to \Delta$ having positive 'minimal displacement' $\inf\{\|x - T_2 x\| : x \in \Delta\}$.

To begin the construction we consider the 'frame' $\mathscr{T}$ of Δ where

$$\mathscr{T} = \bigcup_{n=0}^{\infty} \mathscr{T}_n; \qquad \mathscr{T}_n = \operatorname{conv}\{e_n, e_{n+1}\} \equiv [e_n, e_{n+1}].$$

Since $\mathscr{T}$ is a piecewise linear curve it may be parameterized by a function $g: [0, \infty) \to \mathscr{T}$ defined as follows. For $t \in [0, \infty)$, let $t = n + s$ where n is the integer part of t and $0 \leqslant s < 1$. Set

$$g(t) = g(n + s) = (1 - s)e_n + se_{n+1}.$$

Observe that any point $x \in \Delta$ has a unique representation of the form $x = \lambda g(t)$ with $0 \leqslant \lambda \leqslant 1$. We now define T_2 in terms of the 'coordinates' (λ, t). First, define T_2 on the initial triangle Δ_0 of Δ (i.e., for $0 \leqslant t \leqslant 1$) by setting for $x = \lambda g(t)$,

$$T_2 x = \begin{cases} (1 - 3\lambda t)e_0 & \text{for } \lambda \in [0, \tfrac{1}{3}] \\ (1 - t)e_0 + (3\lambda - 1)t\left(\dfrac{e_0 + e_1}{2}\right), & \text{for } \lambda \in [\tfrac{1}{3}, \tfrac{2}{3}] \\ g(t/2 + (3\lambda - 2)), & \text{for } \lambda \in [\tfrac{2}{3}, 1]. \end{cases}$$

The mapping T_2 maps Δ_0 into $\Delta_0 \cup \Delta_1$, and routine calculations show that it is lipschitzian. Moreover, in view of compactness and the fact that Fix $T = \emptyset$ on Δ_0 we have $\inf\{\|x - Tx\| : x \in \Delta_0\} > 0$.

Now we can extend T_2 to all of Δ by putting for $x \in \bigcup_{n=1}^{\infty} \Delta_n, x = \lambda g(t)$ $(t \geqslant 1)$:

$$Tx = \begin{cases} (1 - 3\lambda)e_0, & \text{for } \lambda \in [0, \tfrac{1}{3}] \\ (3\lambda - 1)g(t - \tfrac{1}{2}), & \text{for } \lambda \in [\tfrac{1}{3}, \tfrac{2}{3}] \\ g((t - \tfrac{1}{2}) + (3\lambda - 2)), & \text{for } \lambda \in [\tfrac{2}{3}, 1]. \end{cases}$$

Then T also is lipschitzian and, since the above definition implies

$$\inf\{\|x - Tx\| : x \in \Delta\} = \inf\{\|x - Tx\| : x \in \Delta_1\} > 0,$$

Fix $T = \emptyset$.

The construction used above will now facilitate a proof of Theorem 19.2.

Proof of Theorem 19.2 Let $B = B(0; 1)$ and $B_r = B(0; r)$ for $r \in (0, 1)$. In view of Theorem 19.3 there exists a lipschitzian mapping $T_1 : B_r \to B_r$ with $\inf\{\|x - T_1 x\| : x \in B_r\} = d_0 > 0$. This mapping can be extended to all of B by setting

$$
Tx = \begin{cases} T_1 x, & \text{for } \|x\| \leq r, \\[2mm] \left(\dfrac{1 - \|x\|}{1 - r}\right) T_1\left(\dfrac{rx}{\|x\|}\right) + \left(\dfrac{\|x\| - r}{1 - r}\right)\dfrac{rx}{\|x\|}, & \text{for } r \leq \|x\| \leq 1. \end{cases}
$$

Observe that $d = \inf\{\|x - Tx\| : x \in B\} > 0$. Also, $T : B \to B$ and moves all points of norm 1 radially inward. Now define the mapping

$$
Sx = x + (2/d)(x - Tx), \qquad x \in B.
$$

Then $\|Sx\| \geq 1$ while, if $\|x\| = 1$, x is mapped radially outward. Moreover, S is lipschitzian, so the mapping R defined by

$$
Rx = Sx/\|Sx\|, \qquad x \in B,
$$

is a lipschitzian retraction of the unit ball onto the unit sphere.

The following theorem completes our development of the fact that Brouwer's Theorem in infinite dimensional Banach spaces fails in a very strong sense.

Theorem 19.4 *In any infinite dimensional Banach space there exists a lipschitzian homotopy which contracts the unit sphere to a point.*

Proof Such a homotopy is clearly given by taking $H(t, x) = R((1 - t)x)$, $t \in [0, 1]$, $x \in S$, where R is the lipschitzian retraction of Theorem 19.2.

The observations of this chapter give rise to several questions, among which are the following: What is the *precise* minimum distance a mapping with a given Lipschitz constant can move all points in its domain? What is the Lipschitz constant of the retraction constructed above? We discuss these questions in the next chapter (where we obtain only partial answers).

   We make one final observation. For any $\epsilon > 0$ the ball $B(0; r)$ in any infinite dimensional Banach space contains a sequence $\{x_n\}$ of points satisfying $\text{dist}(x_{n+1}, \text{span}\{x_1, \ldots, x_n\}) \geq r - \epsilon$. Thus, in view of Lemma 19.3 and the fact that all other steps in the proofs of Theorems 19.2 and Theorem 19.3 were constructive, we conclude:

Theorem 19.5 *There exists a universal constant k_0 such that for any Banach space X there is a lipschitzian retraction R of the unit ball of X onto its boundary with $k(R) \leqslant k_0$.*

20

Minimal displacement

We now expand on the results of the previous chapter, beginning with the so-called 'minimal displacement problem'. This is the problem of determining either precisely, or to within parameters, how *near* certain mappings come to having fixed points. Thus given a set K in a Banach space and a mapping $T: K \to K$, one attempts to obtain an estimate for the quantity

$$\inf\{\|x - Tx\|: x \in K\}.$$

This problem is, of course, meaningful only in situations where it is not already known that T has a fixed point. Its study was initiated in (Goebel, 1973) and, while some further results have been obtained (Franchetti, 1986; Furi and Martelli, 1974; Reich, 1975, 1976a), several major questions remain open.

We shall restrict our attention here to the class of k-lipschitzian mappings $(k \geqslant 1)$. Given a subset K of a Banach space and a constant $k \geqslant 1$, let $\mathscr{L}_K(k)$ denote the family of all k-lipschitzian self-mappings of K:

$$\mathscr{L}_K(k) = \{T \in K^K: k(T) \leqslant k\}.$$

(When K is fixed throughout a discussion, we drop the subscript and write $\mathscr{L}(k)$.)

The following observation is basic.

Lemma 20.1 *For any bounded, closed, convex set $K \subset X$ and any $T \in \mathscr{L}(k)$,*

$$\inf\{\|x - Tx\|: x \in K\} \leqslant r(K)(1 - k^{-1}). \tag{20.1}$$

Proof Let $\epsilon > 0$ and select $z \in K$ such that $K \subset B(z; r(K) + \epsilon)$. Now let $T_\epsilon: K \to K$ be defined by

$$T_\epsilon x = (1 - (k + \epsilon)^{-1})z + (k + \epsilon)^{-1} Tx.$$

Since $T \in \mathscr{L}(k)$, $T_\epsilon \in \mathscr{L}(k(k + \epsilon)^{-1})$; hence T_ϵ is a contraction mapping which has a unique fixed point $x_\epsilon \in K$. Thus

$$\|x_\epsilon - Tx_\epsilon\| = (1 - (k+\epsilon)^{-1})\|z - Tx_\epsilon\|$$
$$\leqslant (1 - (k+\epsilon)^{-1})(r(K) + \epsilon)$$

The conclusion follows upon letting $\epsilon \to 0$.

Similarity arguments make it possible to normalize the size of K and consider only those sets K for which $r(K) = 1$. We do this, and define for such a set $K \subset X$ the function φ_K given by

$$\varphi_K(k) = \sup\{\inf\{\|x - Tx\| : x \in K\} : T \in \mathscr{L}_K(k)\}.$$

Now define φ_X by taking

$$\varphi_X(k) = \sup\{\varphi_K(k) : K \subset X, r(K) = 1\},$$

and, in the special case when $K = B(0; 1)$, set

$$\psi(k) = \psi_X(k) = \varphi_{B(0;1)}(k).$$

Thus, by Lemma 20.1 (for $K \subset X$ with $r(K) = 1$),

$$\varphi_K(k) \leqslant \varphi_X(k) \leqslant 1 - k^{-1}; \tag{20.2}$$

$$\psi(k) \leqslant 1 - k^{-1}. \tag{20.3}$$

As usual, we shall drop the subscripts in (20.2) when either K or X is fixed throughout.

The first natural question to arise is: For which sets or spaces are the estimates (20.1), (20.2), (20.3) sharp?

One fact is already known. Example 17.2 shows that (20.2) is sharp for some subsets of $\mathscr{C}[0, 1]$. Consequently

$$\varphi_{\mathscr{C}[0,1]}(k) = 1 - k^{-1}. \tag{20.4}$$

Actually, one can say more:

Example 20.1 Let B denote the unit ball in the space $\mathscr{C}[-1, 1]$, and for fixed $k > 1$ set

$$\alpha(t) = \begin{cases} -1, & -1 \leqslant t \leqslant -k^{-1} \\ kt, & -k^{-1} \leqslant t \leqslant k^{-1} \\ 1, & k^{-1} \leqslant t \leqslant 1. \end{cases}$$

By combining the construction used in Example 3.4 with that of

Example 17.2 we define $T: B \to B$ by taking

$$(Tx)(t) = \alpha(\max\{-1, \min\{1, x(t) + 2t\}\}).$$

Then $k(T) = k$ while for each $x \in \mathscr{C}[-1, 1]$, $\|x - Tx\| \geqslant 1 - k^{-1}$. Since $\mathscr{C}[-1, 1]$ is isometric to $\mathscr{C}[0, 1]$, this implies

$$\varphi_{\mathscr{C}[0,1]}(k) = \psi_{\mathscr{C}[0,1]}(k) = 1 - k^{-1}. \tag{20.5}$$

The same situation as above arises in the space c_0.

Example 20.2 Let B denote the unit ball in the space $X = c_0$, fix $k > 1$, and for $t \in [-1, 1]$ set $a(t) = \min\{1, k|t|\}$. Now define $T: B \to B$ by taking for $x = (x_1, x_2, \dots) \in B$:

$$Tx = (1, a(x_1), a(x_2), \dots).$$

Then

$$Tx - x = (1 - x_1, a(x_1) - x_2, a(x_2) - x_3, \dots).$$

Observe that $\|Tx - x\| > 1 - k^{-1}$ since the reverse inequality implies $x_i \geqslant k^{-1}$, $i = 1, 2, \dots$, which is a contradiction. Thus

$$\varphi_{c_0}(k) = \psi_{c_0}(k) = 1 - k^{-1}; \tag{20.6}$$

and, moreover, in this example $\inf\{\|x - Tx\|: x \in B\}$ is never attained.

In the space l^1 the situation is different.

Example 20.3 Let B denote the unit ball in l^1 and consider the subset K of B given by

$$K = \overline{\mathrm{conv}}\{e^1, e^2 \dots\} = \left\{x = (x_1, x_2, \dots,): x_i \geqslant 0, \sum_{i=1}^{\infty} x_i = 1\right\}.$$

Obviously $r(K) = \mathrm{diam}(K) = 2$. Now suppose $k > 1$. For any $x \in K$ there is a maximal index i for which $\Sigma_{j=i}^{\infty} x_j > k^{-1}$. Denote this index $i_0 = i_0(x)$ and define $\mu(x) \in [0, 1]$ by the relation

$$\mu(x)x_{i_0} + \sum_{j=i_0+1}^{\infty} x_j = k^{-1}.$$

Now define $T_0: K \to K$ by taking

$$T_0 x = T_0(x_1, x_2, \dots,) = k(0, \dots, 0, \mu(x)x_{i_0}, x_{i_0+1}, \dots,).$$

where 0 appears in the last term i_0 times. It can be shown that $T_0 \in \mathscr{L}(k)$ and that $\|x - T_0 x\| \geqslant 2(1 - k^{-1})$ for each $x \in K$, from which

$$\varphi_{(1/2)K} = \varphi_{l^1}(k) = 1 - k^{-1}. \tag{20.7}$$

We now consider the function ψ in l^1. Suppose $T \in \mathcal{L}(k)$ with $T: B \to B$ (and $k > 1$) and take any $A > k$. Now let $F: B \to B$ be the approximating map given implicitly by

$$Fx = (1 - A^{-1})x + A^{-1}TFx, \qquad x \in B.$$

Observe that $F \in \mathcal{L}[(A-1)/(A-k)]$ and $TF \in \mathcal{L}\{k[(A-1)/(A-k)]\}$. Moreover, if $\|x - Tx\| \geqslant d > 0$ for all $x \in B$, then $\|x - Fx\| \geqslant d/(A-1)$ and $\|x - TFx\| \geqslant Ad/(A-1)$.

Suppose, for the moment, that $TF0$ is finite dimensional, i.e., that $TF0$ lies in a subspace of l^1 spanned by n basis vectors. Then if P_n denotes the natural projection of l^1 onto this subspace, the mapping $P_n TF$ is finite dimensional and by Brouwer's Theorem there is a point $y \in P_\epsilon(B)$ such that $y = P_n TFy$. Thus

$$k\left(\frac{A-1}{A-k}\right)\|y\| \geqslant \|TFy - TF0\|$$
$$= \|TF - y\| + \|y - TF0\|$$
$$\geqslant \|TFy - y\| + \|TF0\| - \|y\|$$
$$\geqslant 2d\left(\frac{A}{A-1}\right) - \|y\|$$

implying

$$\|y\| \geqslant 2\left(\frac{A}{A-1}\right)\left(\frac{A-k}{A(k+1)-2k}\right)d. \tag{20.8}$$

On the other hand,

$$d\left(\frac{A}{A-1}\right) \leqslant \|TFy - y\| \tag{20.9}$$
$$= \|TFy - P_n TFy\|$$
$$= \|TFy\| - \|y\|$$
$$\leqslant 1 - \|y\|.$$

Combining (20.8) and (20.9),

$$d \leqslant \left(\frac{A-1}{A}\right)\left(\frac{A(k+1)-2k}{A(k+3)-4k}\right) \tag{20.10}$$

for any $A > k$.

If $TF0$ is not 'finite dimensional' then (20.10) can be obtained by taking $\epsilon > 0$, choosing n so large that $\| TF0 - P_n TF0 \| < \epsilon$, and repeating the above arguments with ϵ-proximity.

Finally, by choosing $T \in \mathcal{L}(k)$ for which $\inf\{\| x - Tx \| : x \in B\} = d \geqslant \psi(k) - \epsilon$ and minimizing the right side of (20.10) with respect to A, we obtain

$$\psi_{l^1}(k) \leqslant \begin{cases} \dfrac{2 + 3^{1/2}}{4}(1 - k^{-1}) & \text{for } 1 \leqslant k \leqslant 3 + 2(3)^{1/2}, \\[2ex] \dfrac{k+1}{k+3} & \text{for } k > 3 + 2(3)^{1/2}. \end{cases} \tag{20.11}$$

Thus, in contrast to $\mathscr{C}[0,1]$ and c_0, we have

$$\psi_{l^1}(k) < \varphi_{l^1}(k) = 1 - k^{-1}.$$

Before turning to generalities we consider one more example.

Example 20.4 Let H denote a Hilbert space with unit ball B and let K be an arbitrary bounded, closed, convex subset of H. If $T \in \mathcal{L}_K(k)$ then an extension $\tilde{T}$ of T with $\tilde{T} \in \mathcal{L}_H(k)$ is given by $\tilde{T}x = T \circ \mathrm{Proj}_K x$, $x \in H$. Moreover, if $x \notin K$ then $\| x - \tilde{T}x \| \geqslant \| \mathrm{Proj}_K x - T \circ \mathrm{Proj}_K x \|$. By restricting such an extension to the smallest ball containing K we see that $\varphi_H(k) = \psi_H(k)$.

Now let $T \in \mathcal{L}_B(k)$ with $\| x - Tx \| \geqslant d > 0$, and consider the approximating mapping F for T and $A > k > 1$ as in the previous example:

$$Fx = (1 - A^{-1})x + A^{-1}TFx, \qquad x \in B.$$

Then F is strongly monotone; indeed

$$\begin{aligned} k^2 \| Fx - Fy \|^2 &\geqslant \| TFx - TFy \|^2 \\ &= \| AFx - (A-1)x - AFy + (A-1)y \|^2 \\ &= A^2 \| Fx - Fy \|^2 - 2A(A-1)\langle Fx - Fy, x - y \rangle \\ &\quad + (A-1)^2 \| x - y \|^2 \end{aligned}$$

from which

$$2\langle Fx - Fy, x - y \rangle \geqslant \frac{A^2 - k^2}{A(A-1)} \| Fx - Fy \|^2 + \left(\frac{A-1}{A}\right) \| x - y \|^2$$

$$\geqslant \frac{A-1}{A} \| x - y \|^2. \tag{20.12}$$

This in turn implies

$$1 \geqslant \|TF^20\|^2 = \|TF^20 - F0 + F0 - 0\|^2$$
$$= \|TF^20 - F0\|^2 + 2\langle TF^20 - F0, F0 - 0\rangle + \|F0 - 0\|^2$$
$$\geqslant A^2\|F^20 - F0\|^2 + 2A\langle F^20 - F0, F0 - 0\rangle + \|F0 - 0\|^2$$
$$\geqslant A^2\|F^20 - F0\|^2 + (A - 1)\|F0 - 0\|^2 + \|F0 - 0\|^2$$
$$= A^2\|F^20 - F0\|^2 + A\|F0 - 0\|^2$$
$$\geqslant \left(\frac{A^2 + A}{(A - 1)^2}\right)d^2. \tag{20.13}$$

Hence

$$d \leqslant \frac{A - 1}{(A^2 + A)^{1/2}} = (1 - A^{-1})\left(\frac{A}{A + 1}\right)^{1/2},$$

and letting $A \to k$, we obtain the estimate:

$$\varphi_H(k) = \psi_H(k) \leqslant (1 - k^{-1})\left(\frac{k}{k + 1}\right)^{1/2}. \tag{20.14}$$

It is not known whether the estimates (20.11) for l^1 and (20.14) for H are sharp. Also, it is not clear for what spaces X the functions φ_X and ψ_X are minimal. The estimates (20.11) and (20.14) do not suggest an obvious conjecture because (20.14) gives a smaller estimate than (20.11) for k near 1, while for large k the situation is reversed.

We now establish some general properties of the functions φ and ψ. We begin by listing some properties which hold for all such functions.

Lemma 20.2 *If φ denotes any of the functions $\varphi_X, \varphi_K, \psi_X$, then:*
(a) $\varphi(1 - \alpha + \alpha k) \geqslant \alpha\varphi(k), 0 \leqslant \alpha \leqslant 1$;
(b) $\varphi(k)(k - 1)^{-1}$ *is nonincreasing for $k > 1$;*
(c) $\varphi(k)k(k - 1)^{-1}$ *is nondecreasing for $k > 1$;*
(d) $\lim_{k \to 1} \varphi(k)/(k - 1) = \varphi'(1)$ *always exists.*

Proof (a) follows from the simple observation that if $T \in \mathcal{L}_K(k)$ and $\alpha \in [0, 1]$, then

$$T_\alpha = (1 - \alpha)I + \alpha T \in \mathcal{L}(1 - \alpha + \alpha k)$$

with $\|x - T_\alpha x\| = \alpha\|x - Tx\|$ for all $x \in K$. (b) is a reformulation of (a). To prove (c), consider the approximating map for $A > k$ given by

$$Fx = (1 - A^{-1})x + A^{-1}TFx, \qquad x \in K.$$

We have already noted that $F \in \mathcal{L}((A-1)/(A-k))$ and

$$TF \in \mathcal{L}(k(A-1)/(A-k)).$$

Also, if $\|x - Tx\| \geq d > 0$ for all $x \in K$, then

$$\|x - TFx\| = \frac{A}{A-1}\|Fx - TFx\|$$

$$\geq \left(\frac{A}{A-1}\right)d.$$

Thus for all $A > k$,

$$\varphi\left(k\left(\frac{A-1}{A-k}\right)\right) \geq \frac{A}{A-1}\varphi(k). \tag{20.15}$$

Denoting $k(A-1)/(A-k) = l > k$, (20.15) takes the form

$$\frac{\varphi(l)l}{l-1} \geq \frac{\varphi(k)k}{k-1}, \qquad l > k. \tag{20.16}$$

Finally, (d) follows from (b) and (c) (or (20.16)).

We may now formulate the following stronger version of Lemma 20.1.

Theorem 20.1 *For any bounded, convex subset K of a Banach space (with $r(K) = 1$),*

$$\varphi'_K(1)(1 - k^{-1}) \leq \varphi_K(k) \leq 1 - k^{-1}; \tag{20.17}$$

$$\varphi'_X(1)(1 - k^{-1}) \leq \varphi_X(k) \leq 1 - k^{-1}; \tag{20.18}$$

$$\psi'_X(1)(1 - k^{-1}) \leq \psi_X(k) \leq 1 - k^{-1}. \tag{20.19}$$

Observe also that in the notation of this chapter, the result of Lin and Sternfeld (Theorem 19.3) has the following formulation: *For any nonempty, noncompact, bounded, closed and convex subset K of a Banach space and any $k > 1$, $\varphi_K(k) > 0$ (and $\varphi'_K(1) > 0$).*

We now take up the problem of determining minimal values of $\varphi'_K(1)$, $\varphi'_X(1)$ and $\psi'_X(1)$. Spaces X for which $\varphi_X(k) = 1 - k^{-1}$ are precisely those for which $\varphi'_X(1) = 1$. As we have seen, this is the case for the spaces $\mathscr{C}[0, 1]$, l^1 and c_0. However, many other spaces share this property.

For Banach spaces X and Y, we say that Y is *almost isometrically embedded* in X if for any $\epsilon > 0$ there exists a linear operator $A: Y \to X$ such that for all $y \in Y$,

$$(1 - \epsilon)\|y\| \leq \|Ay\| \leq (1 + \epsilon)\|y\|.$$

The following is obvious.

Lemma 20.3. *If Y is almost isometrically embedded in X, then for all* $k > 1$, $\varphi_Y(k) \leqslant \varphi_X(k)$.

It is known (see, e.g., James, 1950; 1964b; see also Lindenstrauss and Tzafriri, 1977) that either c_0 or l^1 is almost isometrically embedded in any nonreflexive space X with an unconditional basis. Thus for all such spaces $\varphi_X(k) = 1 - k^{-1}$ for $k > 1$. The same is true of L^1 since it contains l^1 almost isometrically. However, for spaces having better geometric structure the situation is different. To see this, we shall need the following. (φ is used generically for $\varphi_X, \varphi_K, \psi_X$.)

Lemma 20.4 *For any Banach space X,*

$$2\varphi(2) + \delta_X(2\varphi(2)) \leqslant 1; \tag{20.20}$$

consequently

$$\varphi'(1) + \delta_X(\varphi'(1)) \leqslant 1. \tag{20.21}$$

Proof First we recall (cf., Lemma 10.1) a simple fact about the modulus of convexity. If $\|u\| \leqslant 1$, $\|v\| \leqslant 1$, and $c \in (0, \frac{1}{2})$, then

$$2c\delta(\|u - v\|) \leqslant 2c(1 - (\tfrac{1}{2}))\|u + v\|$$

$$\leqslant 1 - \|cu + (1 - c)v\|.$$

With this the following is immediate: For any $r > 0$, $c \in [0, 1]$, $\epsilon \in [0, 2]$, and $a, x, y \in X$,

$$\left.\begin{array}{l} \|x - a\| \leqslant r \\ \|y - a\| \leqslant r \\ \|x - y\| \geqslant \epsilon r \end{array}\right\} \Rightarrow \|(1 - c)x + cy - a\| \leqslant (1 - 2\min\{c, 1 - c\}\delta(\epsilon))r. \tag{20.22}$$

Now take $\mu > 0$ and let $K \subset X$ be a closed, convex set, with $0 \in K$ and $r(K) = 1$, for which there exists $T \in \mathscr{L}_K(k)$ satisfying

$$\|x - Tx\| \geqslant \varphi(k) - \mu.$$

We may assume, by translating if necessary, that $K \subset B(0; 1 + \mu)$. Take $A = k + \mu$ and define F as usual:

$$Fx = (1 - A^{-1})x + A^{-1}TFx, \qquad x \in K.$$

Set $y = F0$ and consider $\alpha : [0, 1] \to \mathbb{R}$ given by $\alpha(t) = \|TFy - tTF0\|$. Since

α is convex and $\alpha(0) \leqslant 1 + \mu$, and since

$$\alpha(1) = \|TFy - TF0\|$$
$$\leqslant k\|Fy - y\|$$
$$\leqslant (k + \mu)\|Fy - y\|$$
$$= \|TFy - y\|$$
$$= \alpha\left(\frac{1}{k + \mu}\right),$$

we conclude $\alpha(1) = \|TFy - TF0\| \leqslant 1 + \mu$. In (20.22) set $a = TFy$, $x = 0$, $y = TF0$, $r = 1 + \mu$, and $\epsilon = [(k + \mu)(\varphi(k) - \mu)]/[(k + \mu - 1)(1 + \mu)]$. Then

$$\|x - y\| = \|TF0\| = \|TF0 - 0\| \geqslant [A/(A - 1)](\varphi(k) - \mu),$$

so

$$\left(\frac{k + \mu}{k + \mu - 1}\right)(\varphi(k) - \mu)$$

$$\leqslant \|TFy - y\|$$

$$\leqslant \left[1 - 2\min\left\{\frac{1}{k + \mu}, 1 - \frac{1}{k + \mu}\right\}\delta\left(\frac{(k + \mu)(\varphi(k) - \mu)}{(k + \mu - 1)(1 + \mu)}\right)\right](1 + \mu).$$

Letting $\mu \to 0$,

$$\frac{k\varphi(k)}{k - 1} \leqslant 1 - 2\min\left\{\frac{1}{k}, 1 - \frac{1}{k}\right\}\delta\left(\frac{k\varphi(k)}{k - 1}\right),$$

and this establishes (20.20) upon taking $k = 2$. Also, since $(k\varphi(k))/(k - 1)$ increases with k, $(k\varphi(k))/(k - 1) \leqslant 2\varphi(2)$ for $k \in [1, 2]$, so (20.20) implies

$$\frac{k\varphi(k)}{k - 1} + \delta\left(\frac{k\varphi(k)}{k - 1}\right) \leqslant 1.$$

(20.21) now follows upon letting $k \to 1$.

The following is now immediate from Lemma 20.4.

Theorem 20.2 *If X is any Banach space for which $\epsilon_0(X) < 1$, then $\varphi'_X(1) < 1$ and for all $k > 1$, $\varphi_X(k) < 1 - k^{-1}$.*

Very little more is known about the qualitative properties of the φ-functions, although some further properties of these functions arise in the next chapter, which again deals with lipschitzian retractions.

21

The retraction problem

In Chapter 19 we saw that for any infinite dimensional Banach space X there is a lipschitzian retraction R which maps the unit ball B in X onto its unit sphere S. Three questions arise immediately. What is the smallest Lipschitz constant such a retraction might have? How might such a retraction be constructed? Is there a relation between the smallest Lipschitz constant k and the function $\psi_X(k)$? At present only a few preliminary observations are known, and the current methods seem inadequate. Nevertheless we summarize here what is known, beginning with some very basic facts and examples.

Given a Banach space X, let $k_0(X)$ denote the infimum of the set of all numbers $k > 1$ for which there exists a retraction $R \in \mathscr{L}_B(k)$ mapping the unit ball B of X onto its unit sphere S. The final result of Chapter 19 asserts that there is a universal constant k_0 such that $k_0(X) \leqslant k_0$ for any space X.

We begin with the following construction.

Let $R: B \to S$ be a retraction with $R \in \mathscr{L}(k)$, and set $T = -R$. Then $T: B \to S$ and $T^2 = R$. Now let

$$d = \inf\{\|x - Tx\| : x \in B\}$$
$$= \inf\{\|x + Rx\| : x \in B\},$$

and for $\epsilon > 0$ select $x \in B$ such that $\|x - Tx\| \leqslant d + \epsilon$. Now define the curve $\gamma: [0, 1] \to S$ by

$$\gamma(t) = T((1 - t)x + tTx).$$

Then γ is a lipschitzian (thus rectifiable) curve lying in S which joins two antipodal points of S. Let $g(X)$ denote the infimum of the lengths of all such curves. It is obvious that $g(X) \geqslant 2$ for any space X with the inequality strict for some spaces (e.g., $g(H) = \pi$ for a Hilbert space H (Schaffer, 1976)).

Since the curve γ defined above is k-lipschitzian, its length, $l(\gamma)$ satisfies

$$g(X) \leqslant l(\gamma) \leqslant k(d + \epsilon).$$

This, in particular, shows that $d > 0$. Moreover, since $T \in \mathscr{L}(k), d \leqslant \psi(k)$ where $\psi(k) = \psi_X(k)$ is the minimal displacement constant of the previous

219

chapter. This gives the relation

$$k\psi(k) \geqslant g(X). \tag{21.1}$$

Since $\psi(k) \leqslant 1 - k^{-1}$ and $g(X) \geqslant 2$, we obtain the preliminary estimate

$$k_0(X) \geqslant 3 \tag{21.2}$$

for all spaces X.

The estimate (21.2) is likely very imprecise; slightly better estimates are known for certain spaces.

Example 21.1 If $X = l^1$ then $g(X) = 2$, and using (20.11) to solve (21.1) we obtain

$$k(l^1) \geqslant 17 - 8(3^{1/2}) = 3.143\ldots; \tag{21.3}$$

thus $k(l^1)$ is bounded below by a number very slightly larger than π.

Example 21.2 If $X = H$ is a Hilbert space, then by combining (20.14) and (21.1) we have

$$(k-1)\left(\frac{k}{k+1}\right)^{1/2} \geqslant \pi.$$

The solution of the above may be numerically approximated to show that $k_0(H) \geqslant 4.47$. However this can be improved slightly in a straightforward manner. Let $R: B \to S$ be a k-lipschitzian retraction, let $T = -R$, and for $A > k$, set

$$Fx = (1 - A^{-1})x + A^{-1}TFx, \qquad x \in B.$$

Now let $d = \inf\|x - Tx\|$. Utilizing (20.12) as in Example 20.4 we obtain

$$1 = \|TF^2 0\|^2 = \|TF^2 0 - F0 + F0 - 0\|^2$$

$$= \|TF^2 0 - F0\|^2 + 2\langle TF^2 0 - F0, F0 - 0\rangle + \|F0 - 0\|^2$$

$$\geqslant A^2 \|F^2 0 - F0\|^2 + 2A\langle F^2 0 - F0, F0 - 0\rangle + \|F0\|^2$$

$$\geqslant \left(\frac{A}{A-1}\right)^2 \|F^2 0 - TF^2 0\|^2 + (A-1)\|F0\|^2 + \|F0\|^2$$

$$= \left(\frac{A}{A-1}\right)^2 \|F^2 0 - TF^2 0\|^2 + A\|F0\|^2$$

$$\geqslant \left(\frac{A}{A-1}\right)^2 d^2 + \frac{1}{A}.$$

Thus $d \leqslant (1 - A^{-1})^{3/2}$. Since $A > k$ is arbitrary, (21.1) implies

$$k(1 - k^{-1})^{3/2} \geqslant \pi,$$

and this provides a numerical estimate for $k_0(H)$ which is larger than 4.5.

Another consequence of (21.1) is the following estimate for $\psi_X'(1)$. Observe that for any $k > k_0(X)$,

$$\psi_X'(1) \geqslant \frac{\psi(k)}{k-1} \geqslant \frac{g(X)}{k(k-1)}$$

which in turn implies

$$\psi_X'(1) \geqslant \frac{g(X)}{k_0(X)(k_0(X)-1)} \geqslant \frac{2}{k_0(k_0-1)}.$$

Hence all derivatives of ψ at 1 are bounded below by a universal constant.

The above estimate can be easily improved. Let $R \in \mathscr{L}(k)$ be a retraction of B onto S, let $A > k$, and consider the mapping F given by

$$Fx = x - (A^{-1})Rx, \qquad x \in B.$$

Then $F \in \mathscr{L}(1 + (k/A))$, and if J is the normalized duality mapping, for any $x, y \in B$ and $u \in J(x - y)$,

$$\langle Fy - Fx, u \rangle \geqslant \alpha \|y - x\|^2$$

with $\alpha = (1 - (k/A)) > 0$. (Thus F is also strongly accretive.)

The points of S are mapped by F radially inward and $\|x\| = 1 \Rightarrow \|Fx\| = 1 - A^{-1}$. We first claim that $F: B \rightarrow (1 - A^{-1})B$. Indeed, if for $x \in B$, $\|Fx\| > 1 - A^{-1}$, then it is possible to choose $t > 0$ such that $y = x + tFx \in S$. Therefore if $u \in J(tFx) = J(y - x)$,

$$\alpha \|y - x\|^2 = \alpha t^2 \|Fx\|^2 \leqslant \langle Fy - Fx, u \rangle$$
$$= \langle F(x + tFx), u \rangle - \langle Fx, u \rangle$$
$$\leqslant \|F(x + tFx)\| \|u\| - t\|Fx\|^2$$
$$= t\|Fy\| \|Fx\| - t\|Fx\|^2$$

implying

$$\|Fy\| \geqslant \|Fx\|(1 + t\alpha) > 1 - A^{-1},$$

which is a contradiction.

It follows that $F: (1 - A^{-1})B \rightarrow (1 - A^{-1})B$, and since $\|x - Fx\| = A^{-1}$

with $F \in \mathcal{L}(1+(k/A))$,

$$\psi_X\left(1+\frac{k}{A}\right) \geqslant \frac{1}{A}\left(1-\frac{1}{A}\right)^{-1} = \frac{1}{A-1}.$$

Letting $A \to k \geqslant k_0(X)$ we conclude

$$\psi'_X(1) \geqslant \psi(2) \geqslant \frac{1}{k_0(X)-1}.$$

The following simple construction yields this and even more. Let $\epsilon > 0$ and define $T_\epsilon : B \to B$ by taking

$$T_\epsilon x = \begin{cases} -R(x/\epsilon) & \text{for } \|x\| \leqslant \epsilon \\ -x/\|x\| & \text{for } \|x\| \geqslant \epsilon. \end{cases}$$

Thus $T_\epsilon \in \mathcal{L}(k/\epsilon)$ and $\|x - T_\epsilon x\| \geqslant 1 - \epsilon$ for all $x \in B$. Setting $l = k/\epsilon$ we obtain

$$\psi_X(l) \geqslant 1 - k/l,$$

and since k may be chosen arbitrarily near $k_0(X)$, for all $k > k_0$:

$$\psi_X(k) \geqslant 1 - k_0(X)/k \geqslant 1 - k_0/k. \tag{21.4}$$

This combined with the fact that ψ_X is concave at 1 again shows that $\psi'_X(1)$ is always bounded below by a positive number. Moreover, (21.4) provides an estimate for $\psi_X(k)$ for large k.

Since all the above estimates are probably far from sharp, we summarize the qualitative facts:

Theorem 21.1 *If X is any Banach space, then*

(a) $\lim_{k \to \infty} \psi_X(k) = 1.$

Moreover,

(b) $\inf_X \psi'_X(1) > 0.$

Although $\lim_{k \to \infty} \varphi_K(k) \leqslant 1$, the inequality may be strict for sets other than a ball.

Example 21.3 In the Hilbert space l^2 consider the set

$$K = \overline{\text{conv}}\{e^i\} = \left\{ x = \{x_i\} : x_i \geqslant 0, \sum_{i=1}^{\infty} x_i \leqslant 1 \right\}.$$

A technical proof, which we omit, shows that for $k > 1$,

$$\varphi_K(k) \leqslant (1/2^{1/2})(1 - k^{-1});$$

thus in spite of the fact that $r(K) = 1$, $\lim_{k \to \infty} \varphi(k) \leqslant 1/2^{1/2} < 1$.

Note that we have not obtained a value for the infimum over X of $\psi'_X(1)$, nor do we know whether $\psi'_H(1) \leqslant \psi'_X(1)$ even for a Hilbert space H.

We now turn to a consideration of upper bound estimates for the optimal retraction constant $k_0(X)$. The construction of the previous chapter seems too complicated to provide the optimal constant, or to suggest what it might be. Intuitively, one would expect that the value $k_0(X)$ would depend somehow on the geometry of X, and that it should increase as this geometry becomes more structured. We begin with examples of some retractions with relatively small Lipschitz constants.

Examples 21.4 Consider the subspace $\mathscr{C}_0[0, 1]$ of $\mathscr{C}[0, 1]$ consisting of those $f \in \mathscr{C}[0, 1]$ for which $f(0) = 0$. First we show that the unit sphere S in $\mathscr{C}[0, 1]$ is contractible to a point by a lipschitzian homotopy via a combination of the following three homotopies: For $f \in S$ and $c \in [0, 1]$, set

$$H_1(c, f)(t) = \min\{1, |(1 + 2c)f(t) + 1| - 1\};$$

$$H_2(c, f)(t) = \max\{c - 1 + ct, f(t)\};$$

$$H_3(c, f)(t) = (1 - c)f(t) + ct.$$

Note that since $\|f\| = 1$, f must assume the value 1 or -1 (or both). If $f(t) = 1$ for some t then also $H_1(c, f)(t) = 1$. If $f(t) = -1$, then $H_1(c, f)(t) = 2c - 1$ and, in view of the construction of H_1 and the fact that f takes on all values between -1 and 0 on the interval $[0, t]$, there is an $s \in (0, t)$ for which $H_1(c, f)(s) = -1$. Therefore, the homotopy H_1 joins any function $f = H_1(0, f)$ to the function $H_1(1, f)$ which always assumes the value 1.

The second homotopy H_2 applies to a function $f \in S$ which assumes the value 1 and joins f to $H_2(1, f)$ along a path in S. Finally, $H_2(1, f)(t) = \max\{t, f(t)\}$, while H_3 contracts the convex subset of S consisting of functions f for which $f(t) \geqslant t$ to the function $H_3(1, f) \in S$ where $H_3(1, f)(t) \equiv t$.

Now observe that for $f, g \in S$ and $c, d \in [0, 1]$,

$$\|H_1(c, f) - H_1(d, g)\| \leqslant 2|c - d| + 3\|f - g\|;$$

$$\|H_2(c, f) - H_2(d, g)\| \leqslant 2|c - d| + \|f - g\|;$$

and finally, if $f(t) \geqslant t$ and $g(t) \geqslant t$ for all $t \in [0, 1]$,

$$\|H_3(c, f) - H_3(d, g)\| \leqslant |c - d| + \|f - g\|.$$

It is now possible to piece these homotopies together and define the homotopy H as follows:

$$H(c,f) = \begin{cases} H_1(5c/2, f) & \text{for } c \in [0, \tfrac{2}{5}] \\ H_2(5c/2 - 1, H_1(1, f)) & \text{for } c \, [\tfrac{2}{5}, \tfrac{4}{5}]; \\ H_3(5c - 4, H_2(1, H_1(1, f))) & \text{for } c \in [\tfrac{4}{5}, 1]. \end{cases}$$

Thus H joins the functions $f \in S$ to the identity function $H(1, f)$ on $[0, 1]$, and

$$\|H(c,f) - H(d, g)\| \leqslant 5|c - d| + 3\|f - g\|.$$

Now we use H to construct a retraction R of the unit ball B onto S. Let $r \in (0, 1)$ and define

$$Rf = \begin{cases} f_0(t) \equiv t & \text{if } \|f\| \leqslant r \\ H(1 - a(\|f\|), f/\|f\|) & \text{if } \|f\| \geqslant r, \end{cases}$$

where a is any increasing convex function defined on $[r, 1]$ and satisfying $a(r) = 0$, $a(1) = 1$. It is easy to see that R is lipschitzian, but the Lipschitz constant of R depends strongly on the choice of the function a and radius r. To obtain an optimal estimate, let $f, g \in S$ with $\|f\| \geqslant r$, $\|g\| \geqslant r$. Then

$$\|Rf - Rg\| = \|H(1 - a(\|f\|), f/\|f\|) - H(1 - a(\|g\|), g/\|g\|)\|$$

$$\leqslant 5|a(\|f\|) - a(\|g\|)| + 3\|f/\|f\| - g/\|g\|\|$$

$$\leqslant 5a'(\max\{\|f\|, \|g\|\})|\|f\| - \|g\|| + (6/(\max\{\|f\|, \|g\|\}))\|f - g\|$$

$$\leqslant \max_{s \in [r,1]} \{5a'(s) + 6/s\} \|f - g\|$$

In order to make optimal choices for a and r, observe that since

$$\int_r^1 (5a'(t) + 6/t)\, dt = 5 - 6 \ln r$$

does not depend on the function a, the Lipschitz constant of R is minimal if a is chosen to satisfy $5a'(t) + 6/t = (5 - 6 \ln r)/(1 - r) = c(r)$. Then we have

$$\|Rf - Rg\| \leqslant c(r)\|f - g\|. \tag{21.5}$$

Also, c takes on its minimal value when $c(r) = 6/r$ (calculus), and solving for r we obtain $r \simeq 0.345$ and $c(r) \simeq 17.38$. In particular, we have the estimate

$$k_0(\mathscr{C}_0[0, 1]) < 17.38. \tag{21.6}$$

The above construction may be generalized.

Lemma 21.1 *Let X be a Banach space with unit sphere S, and suppose there* *exists a homotopy* $H: [0, 1] \times S \rightarrow S$ *such that for each* $x \in S$, $H(0, x) = x$ *and* $H(1, x) \equiv x_0 \in S$. *Suppose further that constants* A *and* B *exist for which*

$$\|H(c, x) - H(d, y)\| \leqslant A|c - d| + B\|x - y\|$$

for $x, y \in S$, $c, d \in [0, 1]$. *Then there is a retraction* $R: B \rightarrow S$ *satisfying for* $x, y \in B$:

$$\|Rx - Ry\| \leqslant (2B/r)\|x - y\|,$$

where r *is the solution of the equation*

$$2B/r = (A - 2B \ln r)/(1 - r).$$

Proof One need only follow the steps in Example 21.4.

It is doubtful that the above method yields the retraction with smallest Lipschitz constant. However for some spaces it may be simplified slightly by considering homotopies which join the identity map to another simple, but not necessarily constant, map.

Example 21.5 Here we consider the unit sphere S in $L^1(0, 1)$. For any $f \in S$ and any $c \in [0, 1]$, define

$$t_f(c) = \sup\left\{t: \int_0^t |f| \, dt = c\right\}.$$

Set

$$H(c, f) = \begin{cases} |f(t)| & \text{if } t \leqslant t_f(c) \\ f(t) & \text{if } t > t_f(c). \end{cases}$$

Obviously, $H(0, f) = f$ and $H(1, f) = |f|$. Now suppose $f, g \in S$ with $t_f(c) \leqslant t_g(c)$. Then

$$\|H(c, f) - H(c, g)\| = \int_0^{t_f(c)} ||f| - |g|| \, dt + \int_{t_f(c)}^{t_g(c)} |f - |g|| \, dt + \int_{t_g(c)}^1 |f - g| \, dt$$

$$\leqslant \int_0^1 |f - g| \, dt + \int_{t_f(c)}^{t_g(c)} (||f| - |g|| - |f - g|) \, dt$$

$$\leqslant \|f - g\| + 2 \int_{t_f(c)}^{t_g(c)} |g| \, dt. \qquad (21.7)$$

There are two ways to estimate the last integral.

$$\int_{t_f(c)}^{t_g(c)} |g|\, dt = \int_0^{t_g(c)} |g|\, dt - \int_0^{t_f(c)} |g|\, dt$$

$$= c - \int_0^{t_f(c)} |g|\, dt$$

$$= \int_0^{t_f(c)} (|f| - |g|)\, dt$$

$$\leqslant \int_0^{t_f(c)} |f - g|\, dt; \qquad (21.8)$$

$$\int_{t_f(c)}^{t_g(c)} |g|\, dt = \int_{t_f(c)}^1 |g|\, dt - \int_{t_g(c)}^1 |g|\, dt$$

$$= \int_{t_f(c)}^1 |g|\, dt - (1 - c)$$

$$= \int_{t_f(c)}^1 (|g| - |f|)\, dt$$

$$\leqslant \int_{t_f(c)}^1 |f - g|\, dt. \qquad (21.9)$$

Thus

$$\|H(c,f) - H(c,g)\| \leqslant \|f - g\| + 2 \min\left\{ \int_0^{t_f(c)} |f - g|\, dt, \int_{t_f(c)}^1 |f - g|\, dt \right\}$$

$$\leqslant 2\|f - g\|. \qquad (21.10)$$

On the other hand, for any $f \in S$ and $c, d \in [0, 1]$, $c \leqslant d$,

$$\|H(c,f) - H(d,f)\| = \int_{t_f(c)}^{t_f(d)} (|f| - f)\, dt$$

$$\leqslant 2 \int_{t_f(c)}^{t_f(d)} |f|\, dt$$

$$= 2(d - c) \qquad (21.11)$$

In view of (21.10) and (21.11),

$$\|H(c,f)-H(d,g)\| \leqslant 2|c-d|+2\|f-g\|. \tag{21.12}$$

Now let R be the retraction given by

$$Rf = \begin{cases} (1/r)(r-\|f\|+|f|) & \text{if } \|f\| \leqslant r \\ H(1-a(\|f\|), f/\|f\|) & \text{if } \|f\| > r, \end{cases}$$

where the function a is chosen as above (Example 21.4). Then for $\|f\| \leqslant r$ and $\|g\| \leqslant r$,

$$\|Rf-Rg\| \leqslant (2/r)\|f-g\|, \tag{21.13}$$

while the steps of our previous estimate for $\|f\| \geqslant r$, $\|g\| \geqslant r$ lead to

$$\|Rf-Rg\| \leqslant (4/r)\|f-g\|, \tag{21.14}$$

where r is chosen to satisfy

$$\frac{2-4\ln r}{1-r} = \frac{4}{r}. \tag{21.15}$$

Thus $r \simeq 0.424$, and we conclude

$$k_0(L^1(0,1)) < 9.43. \tag{21.16}$$

We know of no example of a space X for which $k_0(X)$ is smaller than the estimate $4/r$ given by (21.15). Also, it should be noted that the homotopy used above is, in some sense, almost optimal. Indeed, for any function $f \leqslant 0$, $H(c,f)$ transforms $[0,1]$ onto a curve joining f to $|f| = -f$, where

$$\|H(c,f)-H(d,f)\| = 2|c-d|.$$

This is a curve of length 2 joining antipodal points of S, and the Lipschitz constant 2 for the parameterization cannot be improved.

Finally, we show what happens with a construction similar to the above in $L^2(0,1)$.

Example 21.6 Here we let B and S denote, respectively, the unit ball and sphere in the Hilbert space $L^2(0,1)$ with standard norm and inner product. As in the previous example, for $f \in S$ and $c \in [0,1]$, set

$$t_f(c) = \sup\left\{t: \int_0^t |f|^2 \, dt = c\right\}$$

and

$$H(c,f) = \begin{cases} |f(t)| & \text{if } t \leqslant t_f(c) \\ f(t) & \text{if } t > t_f(c). \end{cases}$$

Obviously $H(0,f) = f$ and $H(1,f) = |f|$. Moreover, H is uniformly continuous on $[0, 1] \times S$. To see this suppose $f, g \in S$ and $c, d \in [0, 1]$ with $t_f(c) \leqslant t_g(d)$. Then

$$\|H(c,f) - H(d, g)\|^2 = \int_0^{t_f(c)} (|f| - |g|)^2 \, dt + \int_{t_f(c)}^{t_g(d)} |f - |g||^2 \, dt + \int_{t_g(d)}^1 |f - g|^2 \, dt$$

$$\leqslant \int_0^1 |f - g|^2 \, dt + \int_{t_f(c)}^{t_g(d)} (|f - |g||^2 - |f - g|^2) \, dt$$

$$\leqslant \|f - g\|^2 + 4 \int_{t_f(c)}^{t_g(d)} |f| \, |g| \, dt$$

$$\leqslant \|f - g\|^2 + 4 \left(\int_{t_f(c)}^{t_g(d)} |f|^2 \, dt \right)^{1/2} \left(\int_{t_f(c)}^{t_g(d)} |g|^2 \, dt \right)^{1/2}.$$

Each of the last two integrals may be evaluated in two ways as follows:

$$\int_{t_f(c)}^{t_g(d)} |f|^2 \, dt = \int_0^{t_g(d)} |f|^2 \, dt - c$$

$$= \int_0^{t_g(d)} (|f|^2 - |g|^2) \, dt + d - c$$

$$\leqslant |d - c| + \int_0^{t_g(d)} ||f|^2 - |g|^2| \, dt; \qquad (21.18)$$

$$\int_{t_f(c)}^{t_g(d)} |f|^2 \, dt = (1 - c) - \int_{t_g(d)}^1 |f|^2 \, dt$$

$$= (1 - c) - (1 - d) + \int_{t_g(d)}^1 (|g|^2 - |f|^2) \, dt$$

$$\leqslant |d - c| + \int_{t_g(d)}^1 ||g|^2 - |f|^2| \, dt. \qquad (21.19)$$

Combining (21.18) and (21.19) we have

$$\int_{t_f(c)}^{t_g(d)} |f|^2 \, dt \leqslant |c - d| + (\tfrac{1}{2}) \int_0^1 ||f|^2 - |g|^2| \, dt$$

$$\leqslant |c - d| + (\tfrac{1}{2}) \int_0^1 (|f| + |g|)|f - g| \, dt$$

$$\leqslant |c - d| + \|f - g\|.$$

Similarly,

$$\int_{t_f(c)}^{t_g(d)} |g|^2 \, dt \leqslant |c - d| + \|f - g\|.$$

Therefore (21.17) implies

$$\|H(c, f) - H(d, g)\|^2 \leqslant \|f - g\|^2 + 4\|f - g\| + 4|c - d|.$$

This proves that H is uniformly continuous on $[0, 1] \times S$. However (as may be checked) H is *not* lipschitzian. Nevertheless, H may be used to produce a nice uniformly continuous retraction $R: B \to S$.

Let P denote the radial projection of $L^2(0, 1)$ onto its unit ball, and for $f \in B$ and $r \in (0, 1)$, let

$$Rf = \begin{cases} P\left(2^{1/2} \dfrac{r - \|f\| + |f|}{r}\right) & \text{if } \|f\| \leqslant r \\[2ex] H\left(\dfrac{1 - \|f\|}{1 - r}, \dfrac{f}{\|f\|}\right) & \text{if } \|f\| > r. \end{cases} \tag{21.20}$$

The retraction R is uniformly continuous since for $\|f\| \leqslant r$ and $\|g\| \leqslant r$,

$$\|Rf - Rg\| \leqslant \frac{2(2^{1/2})}{r} \|f - g\| \tag{21.21}$$

while if $\|f\| \geqslant r$ and $\|g\| \geqslant r$,

$$\|Rf - Rg\|^2 \leqslant \frac{1}{r^2} \|f - g\|^2 + \frac{4}{r(1 - r)} \|f - g\|. \tag{21.22}$$

The other cases are not difficult to check.

By using Kirzbraun's Theorem it is possible to modify the above retraction to obtain a lipschitzian one which serves to give a rough estimate for $k_0(L^2)$.

For example, consider first the retraction R_0 given by (21.20) when $r = \frac{1}{2}$:

$$R_0 f = \begin{cases} P(2^{1/2}(1 - 2\|f\| + 2|f|)) & \text{if } \|f\| \leq \frac{1}{2} \\ H\left(2(1 - \|f\|), \dfrac{f}{\|f\|}\right) & \text{if } \|f\| \geq \frac{1}{2}. \end{cases}$$

Then (21.21) and (21.22) yield for $f, g \in B(0; \frac{1}{2})$

$$\|R_0 f - R_0 g\| \leq 4(2^{1/2}) \|f - g\|.$$

while for $f, g \in \overline{B(0, 1)} \setminus B(0, \frac{1}{2})$,

$$\|R_0 f - R_0 g\|^2 \leq 4\|f - g\|^2 + 16\|f - g\|.$$

Now take $\epsilon > 0$ sufficiently small $(\epsilon < \frac{1}{4})$ and choose a set $W \subset \overline{B(0; 1 - \epsilon) \setminus B(0; \frac{1}{2} + \epsilon)}$ with the following properties:

(a) For any $x \in B$, dist $(x, B(0; \frac{1}{2}) \cup W \cup S) \leq \epsilon$;
(b) For any $x, y \in W$ with $x \neq y$, $\|x - y\| \geq \epsilon$.

(Use Zorn's lemma to obtain an ϵ-net in $\overline{B(0; 1 - \epsilon) \setminus B(0; (\frac{1}{2}) + \epsilon)}$.)

Now let $V = B(0; \frac{1}{2}) \cup W \cup S$ and let R_1 denote $R_0|_V$. We claim R_1 is lipschitzian on V. Indeed, for $f, g \in B(0; \frac{1}{2})$ we have

$$\|R_1 f - R_1 g\| \leq 4(2^{1/2}) \|f - g\|$$

while for $f, g \in S(0; \frac{1}{2}) \cup W \cup S$.

$$\|R_1 f - R_1 g\|^2 \leq 4\|f - g\|^2 + 16\|f - g\|$$

$$\leq [4 + (16/\epsilon)] \|f - g\|^2;$$

hence

$$\|R_1 f - R_1 g\| \leq (4 + (16/\epsilon))^{1/2} \|f - g\|.$$

On the other hand, if $\|f\| < \frac{1}{2}$ and $g \in W$, select $\alpha \in (0, 1)$ so that $\|h\| = \frac{1}{2}$ where $h = (1 - \alpha)f + \alpha g$. Then

$$\|R_1 f - R_1 g\| \leq \|R_1 f - R_1 h\| + \|R_1 h - R_1 g\|$$

$$\leq 4(2^{1/2}) \alpha \|f - g\| + (4 + (16/\epsilon))^{1/2} (1 - \alpha) \|f - g\|,$$

and since $4(2^{1/2}) \leq (4 + 16/\epsilon)^{1/2}$,

$$\|R_1 f - R_1 g\| \leq (4 + (16/\epsilon))^{1/2} \|f - g\|.$$

By Kirzbraun's Theorem $R_1 : V \to S$ may be extended to a mapping $R_2 : B \to B$ with the same Lipschitz constant. In particular, $R_2 = I$ on S and $R_2 f \neq 0$ for all $f \in B$ if ϵ is chosen sufficiently small. Indeed, for any $f \in B$

there exists $g \in V$ such that $\|f-g\| \leqslant \epsilon$. Thus for $\epsilon > 0$ small:

$$\|R_2 f\| \geqslant \|R_2 g\| - \|R_2 f - R_2 g\|$$

$$\geqslant 1 - (4 + (16/\epsilon))^{1/2} \|f - g\|$$

$$\geqslant 1 - \epsilon(4 + (16/\epsilon))^{1/2} > 0.$$

Finally, set

$$\tilde{R}f = P\left(\frac{R_2 f}{1 - \epsilon(4 + (16/\epsilon))^{1/2}}\right)$$

and observe that for $f, g \in B$,

$$\|\tilde{R}f - \tilde{R}g\| \leqslant \frac{(4 + (16/\epsilon))^{1/2}}{1 - \epsilon(4 + (16/\epsilon))^{1/2}} \|f - g\|.$$

It is now possible to obtain a rough estimate for $k_0(L^2)$ by minimizing the Lipschitz constant for $\tilde{R}$ with respect to ϵ. Simple calculations show that $\epsilon \cong 0.0154$ is almost optimal, and this yields

$$\|\tilde{R}f - \tilde{R}g\| \leqslant 64.25 \|f - g\|.$$

Most of the results of the chapter are preliminary in nature and many appear only in the thesis of Komorowski (1987).

Appendix: Notes and comments

Because of our desire to complete a modest text in a timely fashion, several important topics which fall under the metric fixed point theory umbrella have not been included. Certainly it would have been appropriate to include a treatment of the Nonlinear Mean Ergodic Theorem due to Baillon (1975) and Reich (1979) (also see, e.g., Bruck, 1979). The fixed point theorems for nonlinear semigroups of nonexpansive mappings due to Bruck (1974) and Lim (1974a) also represent fundamental contributions to the theory. Because of their importance, we state these results here.

The Nonlinear Mean Ergodic Theorem *Let K be a bounded, closed, convex subset of a uniformly convex Banach space with Fréchet differentiable norm, and let $T: K \to K$ be nonexpansive. For $x \in K$, let*

$$S_n(x) = (1/n) \sum_{i=0}^{n-1} T^i(x).$$

Then $\{S_n(x)\}$ converges weakly to a fixed point of T.

Bruck's Theorem *Let K be a nonempty, closed, convex subset of a Banach space and suppose K is either weakly compact, or bounded and separable. Suppose also that K has the following property: If $T: K \to K$ is nonexpansive, then T has a fixed point in every nonempty, bounded, closed, convex, T-invariant subset of K. Then for any commuting family S of nonexpansive self-mappings of K, the set $F(S)$ of common fixed points of S is a nonempty, nonexpansive retract of K.*

We remark that the above result is powerful because, in general, the fixed point sets of nonexpansive mappings need not be weakly compact (see Schechtman, 1982).

Lim's version of the above result is formulated for topological semi-groups S which are not only commutative, but more generally, left reversible. (A semigroup S is left reversible if any two closed right ideals have nonempty intersection.) The following is the discrete version of Lim's result.

Lim's Theorem *Let K be a nonempty, weakly compact, convex subset of a Banach space, and suppose K has normal structure. Suppose S is a left reversible discrete semigroup of nonexpansive self-mappings of K. Then the commom fixed point set $F(S)$ of S is nonempty.*

We also point out that in an attempt to emphasize the metric flavor of the subject we have omitted a large body of the theory which is primarily topological in nature, as well as many results which bridge the metric and topological theories. An excellent reference for the topological theory is the recent book of Dugundji and Granas (1982). Results which bridge the two theories include the study of the condensing mappings and k-set contractions in which the pioneering degree-theoretic approaches of Browder, Petryshyn, Nussbaum (e.g., Browder and Nussbaum, 1968; Browder and Petryshyn, 1969; Nussbaum, 1971a,b, 1972,a,b; Petryshyn, 1970, 1972; Petryshyn, and Fitzpatrick, 1974) play a fundamental role. Fortunately, most of this theory is readily accessible elsewhere; for example, excellent expositions of the degree-theoretic approach and its applications are given in Browder (1976) and Deimling (1985). Also, see Browder, 1980.

References

Alspach, D. E. (1981) A fixed point free nonexpansive map, *Proc. Amer. Math. Soc.* **82**, 423–4.

Aronszajn, A. and Panitchpakdi, P. (1956) Extensions of uniformly continuous transformations and hyperconvex metric spaces, *Pacific J. Math.* **6**, 405–39.

Aubin, J. P. and Cellina, A. (1984) *Differential Inclusions*, Springer, Berlin.

Baillon, J. B. (1975) Un théorème de type ergodic pour les contractions nonlinéaires dans un espace de Hilbert, *C. R. Acad. Sci. Paris* **280**, 1511–14.

Baillon, J. P. (1978–79) Quelques aspects de la théorie des points fixes dans les espaces de Banach I, *Séminaire analyse fonctionnelle, Ecole Polytechnique.*

Baillon, J. B. (1988) Nonexpansive mappings and hyperconvex spaces, *Fixed Point Theory and its Applications*, (R. F. Brown, ed.), *Contemporary Mathematics*, vol. 72, American Mathematical Society, Providence, RI.

Baillon, J. B., Bruck, R. and Reich, S. (1978) On the asymptotic behavior of nonexpansive mappings, *Houston J. Math.* **4**, 1–9.

Banach, S. (1922) Sur les opérations dans les ensembles abstraits et leurs applications, *Fund. Math.* **3**, 133–181.

Banas, J. and Goebel, K. (1980) *Measures of Noncompactness in Banach Spaces*, Marcel Dekker, New York and Basel.

Belluce, L. P. (1967–68) Nonexpansive mappings in Banach spaces, *Topics in Functional Analysis*, Univ. British Columbia Lecture Notes.

Belluce, L. P. and Kirk, W. A. (1985) Developments in fixed point theory for nonexpansive mappings, *Trends in the Theory and Practice of Nonlinear Analysis* (V. Lakshmikantham, ed.), North Holland, Amsterdam, pp. 55–61.

Benyamini, Y. and Sternfeld, Y. (1983) Spheres in infinite dimensional normed spaces are Lipschitz contractible, *Proc. Amer. Math. Soc.* **88**, 439–45.

Bessaga, C. (1966) Every infinite dimensional Hilbert space is diffeomorphic with its unit sphere, *Bull. Acad. Polon. Sci.* **14**, 27–31.

Bessaga, C. and Pelczynski, A. (1957) *Selected Topics in Infinite-dimensional Topology*, Monografie Matematyczne, vol. 58, p. 353.

Bielecki, A. (1956) Une remarque sur la méthode de Banach–Cacciopoli–Tikhonov dans la théorie des équations differentielles ordinaires, *Bull. Acad. Polon. Sci.*, Cl. III, **4**, 261–4.

Blumenthal, L. M. (1953) *Theory and Applications of Distance Geometry*, Oxford Univ, Press. London.

Bohnenblust, H. F. and Karlin, S. (1950) On a theorem of Ville, *Contributions to the Theory of Games*, vol. 1 (H. W. and A. W. Tucker, eds.), Princeton Univ. Press, Princeton, NJ, pp. 155–60.

Borwein, J. M. and Sims, B. (1984) Nonexpansive mappings on Banach lattices and related topics, *Houston J. Math.* **10**, 339–56.

Brezis, H. (1973) *Opérateurs maximaux monotone, Lecture Notes*, Vol. 5, North Holland, Amsterdam.

Brezis, H. and Browder, F. (1976) A general principle on ordered sets in nonlinear functional analysis, *Advances in Math.* **21**, 355–64.

Brodskii, M. S. and Milman, D. P. (1948) On the center of a convex set, *Dokl. Akad. Nauk SSSR* **59**, 837–40.

Brouwer, L. E. J. (1912) Über Abbildungen vom Mannigfaltigkeiten, *Math. Ann.* **71**, 97–115.

Browder, F. E. (1965a) Fixed point theorems for noncompact mappings in Hilbert space, *Proc. Nat. Acad. Sci. USA* **43**, 1272–6.

Browder, F.E. (1965b) Nonexpansive nonlinear operators in a Banach space, *Proc. Nat. Acad. Sci. USA* **54**, 1041–4.

Browder, F. E. (1968) Semicontractive and semiaccretive nonlinear mappings in Banach spaces, *Bull. Amer. Math. Soc.* **74**, 660–5.

Browder, F. E. (1976) Nonlinear operators and nonlinear equations of evolution in Banach spaces, *Proc. Symp. Pure Math.*, vol. 18, pt 2, American Mathematical Society, Providence, RI.

Browder, F. E. (1980) Fixed point theory and nonlinear problems, *The Mathematical Heritage of Henri Poincaré* (F. Browder, ed.), *Proc. Symp. Pure Math.* 39, pt 2, American Mathematical Society, Providence, RI, pp. 49–87.

Browder, F. E. and Nussbaum, R. (1968) The topological degree for noncompact nonlinear mappings in Banach spaces, *Bull. Amer. Math. Soc.* **74**, 671–6.

Browder, F. E. and Petryshyn, W. (1969) Approximation methods and the generalized topological degree for nonlinear mappings in Banach spaces, *J. Funct. Anal.* **3**, 217–45.

Bruck, R. E. (1973) Properties of fixed point sets of nonexpansive mappings in Banach spaces, *Trans. Amer. Math. Soc.* **179**, 251–62.

Bruck, R. E. (1974) A common fixed point theorem for a commuting family of nonexpansive mappings, *Pacific J. Math.* **53**, 59–71.

Bruck, R. E. (1979) A simple proof of the mean ergodic theorem for nonlinear contractions in Banach spaces, *Israel J. Math.* **32**, 279–82.

Bruck, R. E. (1981) On the convex approximation property and the asymptotic behavior of nonlinear contractions in Banach spaces, *Israel J. Math.* **38**, 304–14.

Bruck, R. and Reich, S. (1981) Accretive operators, Banach limits and dual ergodic theorems, *Bull. Acad. Polon. Sci.* **29**, 585–9.

Bynum, W. L. (1972) A class of Banach spaces lacking normal structure, *Compositio Math.* **25**, 233–6.

Bynum, W. L. (1980) Normal structure coefficients for Banach spaces, *Pacific J. Math.* 86, 427–36.

Caristi, J. (1976) Fixed point theorems for mappings satisfying inwardness conditions, *Trans. Amer. Math. Soc.* **215**, 241–51.

Clarkson, J. A. (1936) Uniformly convex spaces, *Trans. Amer. Math. Soc.* **40**, 396–414.

Crandall, M. G. and Liggett T. M. (1971) Generation of semigroups of nonlinear transformations on general Banach spaces, *Amer. J. Math.* **93**, 265–98.

Crandall, M. G. and Pazy, A. (1972) Nonlinear evolution equations in Banach spaces, *Israel J. Math.* **11**, 57–94.

Darbo, G. (1955) Punti uniti in transformazioni a codominio non compatto, *Rend. Sem. Mat. Univ. Padova* **25**, 84–92.

Day, M, M. (1941) Reflexive Banach spaces not isomorphic to uniformly convex spaces, *Bull. Amer. Math. Soc.* **47**, 313–17.

Day, M. M. (1973) *Normed Linear Spaces*, 3rd edn, Springer, Berlin, Heidelberg, New York.

Day, M. M., James, R. C. and Swaminathan, S. (1971) Normed linear spaces that are uniformly convex in every direction, *Can. J. Math.* **23**, 1051–9.

Deimling, K. (1985) *Nonlinear Functional Analysis*, Springer, Berlin, Heidelberg, New York, Tokyo.

Diestel, J. (1975) *Geometry of Banach Spaces – Selected Topics, Lecture Notes in Mathematics*, no. 485, Springer, Berlin, New York.

Downing, D. and Ray, W. (1981) Some remarks on set valued mappings, *Nonlinear Analysis* **5**, 1367–77.

Downing, D. and Turett, B. (1983) Some properties of the characteristic of convexity relating to fixed point theory, *Pacific J. Math.* **104**, 343–50.

Dugundji, J. and Granas, A. (1982) *Fixed Point Theory, Monografie Matematyczne*, vol. 61, Polish Scientific Publishers, Warsaw.

Dulst, D., van, (1978) *Reflexive and Supperreflexive Banach Spaces, Mathematical Centre Tracts*, vol. 102, Mathematische Centrum, Amsterdam.

Dulst, D. van (1984) Some more Banach spaces with normal structure, *J. Math. Anal. Appl.* **104**, 285–92.

Dulst, D. van and Sims, B. (1983) Fixed points of nonexpansive mappings and Chebyshev centers in Banach spaces with norms of type (KK), *Banach Space Theory and Its Applications, Lecture Notes in Mathematics*, no. 991, Springer, Berlin, New York.

Dunford, N. and Schwartz, J. (1957) *Linear Operators, Part I: General Theory*, Wiley Interscience, New York.

Dvoretsky, A. (1961) Some results on convex bodies and Banach spaces, *Proc. Int. Symp. Linear Spaces*, Jerusalem, 1960, Jerusalem Academic Press, Jerusalem; pp. 123–60.

Edelstein, M. (1962) On fixed and periodic points under contractive mappings, *J. London Math. Soc.* **37**, 74–9.

Edelstein, M. (1964) On non-expansive mappings of Banach spaces, *Proc. Camb. Phil. Soc.* **60**, 439–47.

Edelstein, M. (1966) A remark on a theorem of M. A. Krasnoselskii, *Amer. Math. Monthly*, **13**, 509–10.

Edelstein, M. (1972) The construction of an asymptotic center with a fixed point property, *Bull. Amer. Math. Soc.* **78**, 206–8.

Edelstein, M. and O'Brien, R. C. (1978) Nonexpansive mappings, asymptotic regularity, and successive approximations, *J. London Math. Soc.* **17**, 547–54.

Ekeland, I. (1974) On the variational principle, *J. Math. Anal. Appl.* **47**, 324–53.

Elton, J., Lin, P. K., Odell, E. and Szarek, S. (1983) Remarks on the fixed point problem for nonexpansive maps, *Fixed Points and Nonexpansive Mappings*, (R. Sine, ed.), *Contemporary Mathematics*, vol. 18, American Mathematical Society, Providence, RI pp. 87–120.

Enflo, P. (1972) Banach spaces which can be given an equivalent uniformly convex norm, *Israel J. Math.* **13**, 281–8.

Fan, Ky (1960/61) A generalization of Tychonoff's fixed point theorem, *Math. Ann.* **142**, 305–10.

Franchetti, C. (1986) Lipschitz maps and the geometry of the unit ball in normed spaces, *Arch. Math.* **46**, 76–84.

Fuchssteiner, B. (1977) Iterations and fixpoints, *Pacific J. Math.* **67**, 73–80.

Furi, M. and Martelli, M. (1974) On the minimal displacement of points under alpha-Lipschitz maps in normed spaces, *Boll. Un. Mat. Ital.* **9**(1975), 791–9.

Genel, A. and Lindenstrauss, J. (1975) An example concerning fixed points, *Israel J. Math.* **22**, 81–6.

Gillespie, A. and Williams, B. (1979) Fixed point theorem for nonexpansive mappings on Banach spaces with uniformly normal structure, *Applicable Anal.* **9**, 121–4.

Goebel, K. (1969) An elementary proof of the fixed point theorem of Browder and Kirk, *Michigan Math. J.* **16**, 381–3.

Goebel, K. (1970) Convexity of balls and fixed point theorems for mappings with a nonexpansive square, *Compositio Math.* **22**, 269–74.

Goebel, K. (1973) On the minimal displacement of points under lipschitzian mappings, *Pacific J. Math.* **48**, 151–63.

Goebel, K. (1975a) On a fixed point theorem for multivalued nonexpansive mappings, *Annal. Univ. Marie Curie-Sklodowska* **29**, 70–72.

Goebel, K. (1975b) On the structure of the minimal invariant sets for nonexpansive mappings, *Annal. Univ. Marie Curie-Sklodowska* **29**, 73–7.

Goebel, K. and Kirk, W.A. (1973) A fixed point theorem for transformations whose iterates have uniform Lipschitz constant, *Studia Math.* **47**, 135–40.

Goebel, K. and Kirk, W. A., (1983) Iteration processes for nonexpansive mappings,

Topological Methods in Nonlinear Functional Analysis (S. P. Singh and S. Thomeier, eds.), *Contemporary Mathematics*, vol. 21, American Mathematical Society, Providence, RI, pp. 115–23.

Goebel, K. and Koter, M. (1981a) A remark on nonexpansive mappings, *Canadian Math. Bull.* **24**, 113–15.

Goebel, K. and Koter, M. (1981b) Fixed points of rotative lipschitzian mappings, *Rend. Sem. Mat. Fis. Milano* **51**, 145–56.

Goebel, K. and Kuczumow, T. (1978a) Irregular convex sets with the fixed point property for nonexpansive mappings, *Colloq. Math.* **40**, 259–64.

Goebel, K. and Kuczumow, T. (1978b) A contribution to the theory of nonexpansive mappings, *Bull. Calcutta Math. Soc.* **70**, 355–7.

Goebel, K. and Massa, S. (1986) Some remarks on nonexpansive mappings in Hilbert spaces, *Nonlinear Functional Analysis and Its Application* (S. P. Singh, V. M. Sehgal and J. Burry, eds.), *MSRI-Korea Publications*, no. 1, pp. 7–11.

Goebel, K. and Reich, S. (1984) *Uniform Convexity, Hyperbolic Geometry, and Nonexpansive Mappings*, Marcel Dekker, New York.

Goebel, K. and Schöneberg, R. (1977) Moon, bridges, birds ... and nonexpansive mappings in Hilbert space, *Bull. Austral. Math. Soc.* **17**, 463–6.

Goebel, K. and Sekowski, T. (1984) The modulus of noncompact convexity, *Annal. Univ. Marie Curie-Sklodowska* **38**, 41–8.

Goebel, K. and Zlotkiewicz, E. (1971) Some fixed point theorems in Banach spaces, *Colloquium Math.* **23**, 103–6.

Goebel, K., Kirk, W. A. and Thele, R. L. (1974) Uniformly lipschitzian families of transformations in Banach spaces, *Canadian J. Math.* **26**, 1245–56.

Göhde, D. (1965) Zum Prinzip der kontraktiven Abbildung, *Math. Nach.* **30**, 251–8.

Gossez, J. P. and Lami Dozo, E. (1969) Structure normale et base de Schauder, *Bull. Acad. Roy. Belgique* **15**, 637–81.

Gossez, J. P. and Lami Dozo, E. (1972) Some geometric properties related to the fixed point theory for nonexpansive mappings, *Pacific J. Math.* **40**, 565–73.

Hanner, O. (1956) On the uniform convexity of L^p and l^p, *Ark. Mat.* **3**, 239–44.

Heinrich, H. (1980) Ultraproducts in Banach space theory, *J. Reine Angew. Math.* **313**, 72–104.

Hirano, N. (1980) A proof of the mean ergodic theorem for nonexpansive mappings in Banach spaces, *Proc. Amer. Math. Soc.* **78**, 361–5.

Huff, R. E. (1980) Banach spaces which are nearly uniformly convex, *Rocky Mountain J. Math.* **10**, 743–49.

Ishikawa, S. (1976) Fixed points and iteration of a nonexpansive mapping in a Banach space, *Proc. Amer. Math. Soc.* **59**, 65–71.

Istratescu, V. I. (1981) *Fixed Point Theory*, D. Reidel, Dordrecht.

James, R. C. (1950) Bases and reflexivity of Banach spaces, *Ann. Math.* **52**, 518–27.

James, R. C. (1957) Reflexivity and the supremum of linear functionals, *Ann. Math.* **66**, 159–69.

James, R. C. (1964a) Weak compactness and reflexivity, *Israel J. Math.* **2**, 101–19.

James, R. C. (1964b) Uniformly non-square Banach spaces, *Ann. Math.* **80**, 542–50.

James, R. C. (1972) Super-reflexive Banach spaces, *Can. J. Math.* **24**, 896–904.

Kakutani, S. (1938) Two fixed point theorems concerning bicompact sets, *Proc. Imp. Acad. Tokyo* **14**, 242–5.

Kakutani, S. (1941) A generalization of Brouwer's fixed point theorem, *Duke Math. J.* **8**, 457–9.

Kakutani, S. (1943) Topological properties of the unit sphere of a Hilbert space, *Proc. Imp. Acad. Tokyo.* **19**, 269–71.

Karlovitz, L. (1976) Existence of fixed points for nonexpansive mappings in spaces without normal structure, *Pacific J. Math.* **66**, 153–6.

Kartsatos, A. G. (1978) On the equation $Tx = y$ in Banach spaces with weakly continuous duality maps, *Nonlinear Equations in Abstract Spaces* (V. Lakshmikantham, ed.) Academic Press, New York.

Topics in metric fixed point theory

Kartsatos, A. G. (1981) Mapping theorems for accretive operators in Banach spaces, *J. Math. Anal. Appl.* **82**, 169–83.

Kato, T. (1967) Nonlinear semigroups and evolution equations, *J. Math. Soc. Japan*, **19**, 508–20.

Kelley, J. L. (1952) Banach spac~s with the extension property, *Trans. Amer. Math. Soc.* **72**, 323–6.

Kelley, J. L. (1965) *General Topology*, D. Van Nostrand, New York.

Khamsi, M. A. (1987) Étude de la propriété du point fixe dans les éspaces de Banach et les éspaces métriques, Thèse de Doctorat de L'Université Paris VI.

Khamsi, M. A., Kozlowski, W. M. and Reich, S. (1989) Fixed point theory in modular function spaces, *Nonlinear Analysis* (to appear).

Khamsi, M. A. and Thurpin, Ph. (1988) Fixed points of nonexpansive mappings in Banach lattices, *Proc. Amer. Math. Soc.* **104**, 1–8.

Kirk, W. A. (1965) A fixed point theorem for mappings which do not increase distances, *Amer. Math. Monthly* **72**, 1004–6.

Kirk, W. A. (1971) A fixed point theorem for mappings with a nonexpansive iterate, *Proc. Amer. Math. Soc.* **29**, 294–8.

Kirk, W. A. (1976) Caristi's fixed point theorem and metric convexity, *Colloquium Math.* **36**, 81–6.

Kirk, W. A. (1981a) Fixed point theory for nonexpansive mappings, *Fixed Point Theory* (E. Fadell and G. Fournier, eds.), *Lecture Notes in Mathematics*, no. 886, Springer, Berlin, New York, pp. 484–505.

Kirk, W. A. (1981b) Nonexpansive mappings in metric and Banach spaces, *Rend. Sem. Mat. Fis. Milano* **61**, 133–44.

Kirk, W. A. (1983) Fixed point theory for nonexpansive mappings II, *Fixed Points and Nonexpansive Mappings* (R. C. Sine, ed.), *Contemporary Mathematics*, vol. 18; American Mathematical Society, Providence, RI, 1p. 121–40.

Kirk, W. A. (1986) Nonexpansive mappings in product spaces, set-valued mappings, and k-uniform rotundity, *Nonlinear Functional Analysis and its Applications* (F. E. Browder, ed.), *Amer. Math. Soc. Symp. Pure Math.*, vol. **45**, 51–64.

Kirk, W. A. (1989) An iteration process for nonexpansive mappings with applications to fixed point theory in product spaces, *Proc. Amer. Math. Soc.* **107**, 411–15.

Kirk, W. A., and Martinez, Yanez C. (1988) Nonexpansive and locally nonexpansive mappings in product spaces, *Nonlinear Analysis* **12**, 719–25.

Kirk, W. A. and Martinez, Yanez, C. (1990) Approximate fixed points for nonexpansive mappings in uniformly convex spaces, *Annal. Polonici Math.*

Kirk, W. A. and Massa, S. (1987) Remarks on asymptotic and Chebyshev centers, (preprint).

Kirk, W. A. and Schöneberg, R. (1980) Zeros of m-accretive operators in Banach spaces, *Israel J. Math.* **35**, 1–8.

Kirk, W. A. and Sternfeld, Y, (1984) The fixed point property for nonexpansive mappings in certain product spaces, *Houston J. Math.* **10**, 207–14.

Kirzbraun, M. D. (1934) Über die Zussamenziehende und Lipschistsche Transformationen, *Fund. Math.* **22**, 77–108.

Klee, V. (1953) Convex bodies and periodic homeomorphisms in Hilbert spaces, *Trans. Amer. Math. Soc.* **74**, 10–43.

Klee, V. (1955) Some topological properties of convex sets, *Trans. Amer. Math. Soc.* **78**, 30–45.

Komorowski, T. (1987) Selected Topics on Lipschitzian Mappings (in Polish), Thesis, Univ. Marii Curie-Sklodowska.

Koter, M. (1986) Fixed points of Lipschitzian 2-rotative mappings, *Boll. Un. Mat. Ital.* Ser. VI, **5**, 321–39.

Krasnoselskii, M. A. (1955) Two observations about the method of successive approximations, *Uspehi Mat. Nauk.* **10**, 123–7.

Kuratowski, K. (1930) Sur les éspaces complets, *Fund. Math.* **15**, 301–9.

Lacey, H. E. (1974) *The Isometric Theory of Classical Banach Spaces*, Springer, Berlin, Heidelberg, New York.

Lami Dozo, E. (1973) Multivalued nonexpansive mappings and Opial's condition, *Proc. Amer. Math. Soc.* **38**, 286–292.

Landes, T. (1984) Permanence properties of normal structure, *Pacific J. Math.* **100**, 125–43.

Landes, T. (1986) Normal structure and the sum property, *Pacific J. Math.* **123**, 127–47.

Lange, H. (1973) Abbildungssatze fur monotone Operatoren in Hilbert- und Banach Raumen, Dissert., Albert-Ludwig Universitat, Freiburg, Germany.

Lifshitz, E. A. (1975) Fixed point theorems for operators in strongly convex spaces, *Voronez Gos. Univ. Trudy Mat. Fak.* **16**, 23–8. (in Russian).

Lim, T. C. (1974a) Characterizations of normal structure, *Proc. Amer. Math. Soc.* **43**, 313–19.

Lim, T. C. (1974b) A fixed point theorem for multivalued nonexpansive mappings in a uniformly convex Banach space, *Bull. Amer. Math. Soc.* **80**, 1123–6.

Lim. T. C. (1980) Asymptotic centers and nonexpansive mappings in some conjugate spaces, *Pacific J. Math.* **90**, 135–43.

Lim. T. C. (1985) On fixed point stability for set-valued contractive mappings with applications to generalized differential equations, *J. Math. Anal. Appl.* **110**, 436–41.

Lim. T. C. (1986) A note on the Clarkson inequality and the Hanner inequality, *Operator Equations and Fixed Point Theorems* (S. P. Singh, V. M. Sehgal, and J. Burry, eds.), *The MSRI-Korea Publications*, no. 1, pp. 51–4.

Lin, P. K. (1985) Unconditional bases and fixed points of nonexpansive mappings, *Pacific J. Math.* **116**, 69–76,

Lin, M. and Sine, R. (1988) Semigroups and retractions in hyperconvex spaces, preprint.

Lin, Bor-Luh, Troyanski, S. L. and Zhang, Wenyao (1989) On the points of continuity and the points of sequential continuity, preprint.

Lin, P. K. and Sternfeld, Y. (1985) Convex sets with the Lipschitz fixed point property are compact, *Proc. Amer. Math. Soc.* **93**, 633–9.

Lindenstrauss, J. and Tzafriri, L. (1977, 1979) *Classical Banach Spaces*, vols. I and II, Springer, Berlin, Heidelberg, New York.

Maluta, E. (1989) Uniformly normal structure and related coefficients, *Pacific J. Math.*

Markin, J. (1968) A fixed point theorem for set valued mappings, *Bull. Amer. Math. Soc.* **74**, 639–40.

Martin, R. H. (1973) Differential equations on closed subsets of a Banach space, *Trans. Amer. Math. Soc.* **179**, 399–414.

Martin, R. H. (1976) *Nonlinear Operators and Differential Equations in Banach Spaces*, John Wiley & Sons, New York.

Mauldin, D. (ed.) (1981) *The Scottish Book: Mathematical Problems from the Scottish Cafe*, Birkhauser, Boston.

Maurey, B. (1981) Points fixes des contractions *sur un convex formé* de L^1, *Seminaire d'Analyse Fonctionnelle 1980–81*, Expose no. VIII, Ecole Polytechnique, Palaiseau.

Menger, K. (1928) Untersuchungen uber allgemeine Metrik, *Math. Ann.* **100**, 75–163.

Michael, E. (1956a) Continuous Selections I, *Ann. Math*, **63**, 361–82.

Michael, E. (1956b) Continuous Selections II, *Ann. Math.* **64**, 562–80.

Michael, E. (1957) Continuous selections III, *Ann. Math.* **65**, 375–90.

Milnor, J. (1978) Analytic proof of the 'Hairy Ball Theorem' and the Brouwer Fixed Point Theorem, *Amer. Math. Monthly* **85**, 521–4.

Nachbin, L. (1950) A theorem of the Hahn–Banach type for linear transformations, *Trans. Amer. Math. Soc.* **68**, 28–46.

Nadler, S. (1969) Multivalued contraction mappings, *Pacific J. Math.* **30**, 475–88.

Nordlander, G. (1960) The modulus of convexity in normed linear spaces, *Ark. Mat.* **4**, 15–17.

Nowak, B. (1979) On the Lipschitz retraction of the unit ball in infinite dimensional Banach spaces onto boundary, *Bull. Acad. Polon. Sci.* **27**, 861–4.

Nussbaum, R. (1971a) Some fixed point theorems, *Bull. Amer. Math. Soc.* **77**, 360–5.

Nussbaum, R. (1971b) The fixed point index for local condensing maps, *Ann. Mat. Pura Appl.* **89**, 217–58.

Nussbaum, R. (1972a) Degree theory for local condensing maps, *J. Math. Anal. Appl.* **37**, 741–66.

Nussbaum, R. (1972b) Some asymptotic fixed point theorems, *Trans. Amer. Math. Soc.* **171**, 349–75.

Opial, Z. (1967a) Weak convergence of the sequence of successive approximations for nonexpansive mappings, *Bull. Amer. Math. Soc.* **73**, 591–7.

Opial, Z. (1967b) *Nonexpansive and Monotone Mappings in Banach Spaces, Lecture Notes* 67–1, Center for Dynamical Systems, Brown University, Providence, RI.

Pasicki, L. (1978) A short proof of the Caristi theorem, *Ann. Soc. Math. Polon. Series I: Comm. Math.* **22**, 427–8.

Peitgen, H-O. and Richter, P. H. (1986) *The Beauty of Fractals: Images of Complex Dynamical Systems*, Springer, Berlin, New York.

Penot, J. (1979) Fixed point theorems without convexity, *Bull. Soc. Math. France, Memoire* **60**, 129–52.

Petryshyn, W. (1970) Structure of the fixed point sets of k-set contractions, *Arch. Rat. Mech. Anal. Appl.* **40**, 312–28.

Petryshyn, W. (1972) Remarks on condensing and k-set contractive mappings, *J. Math. Anal. Appl.* **39**, 717–41.

Petryshyn, W. and Fitzpatrick, P. M. (1974) A degree theory, fixed point theorems and mapping theorems for multivalued noncompact mappings, *Trans. Amer. Math. Soc.* **194**, 1–25.

Prus, S. (1988) A remark on a theorem of Turett (preprint).

Ray, W. O. (1980) The fixed point property and unbounded sets in Hilbert space, *Trans. Amer. Math. Soc.* **258**, 531–7.

Ray, W. O. (1981) Fixed point theory for nonexpansive mappings Seminar notes (unpublished).

Reich, S. (1975) Minimal displacement of points under weakly inward pseudo-lipschitzian mappings I, *Atti. Acad. Naz. Linzei Rend. Cl. Sci. Fis. Mat. Natur.* **59**, 40–4.

Reich, S. (1976a) Minimal displacement of points under weakly inward pseudo lipschitzian mappings II, *Atti. Acad. Naz. Linzei Rend. Cl. Sci. Fis. Mat. Natur.* **60**, 95–6.

Reich, S. (1976b) On fixed point theorems obtained from existence theorems for differential equations, *J. Math. Anal. Appl.* **54**, 26–36.

Reich, S. (1979) Weak convergence theorems for nonexpansive mappings in Banach spaces, *J. Math. Anal. Appl.* **67**, 274–6.

Reich, S. (1980) Strong convergence theorems for resolvents of accretive operators in Banach spaces, *J. Math. Anal. Appl.* **75**, 287–92.

Reich, S. (1983) The almost fixed point property for nonexpansive mappings, *Proc. Amer. Math. Soc.* **88**, 44–6.

Ricceri, B. (1987) Structure, approximation et dependance continue des solutions de certaines equations non lineaires, *C. R. Acad. Sci. Paris* **305** (Serie I), 45–7.

Ricceri, B. (1988) Smooth extensions of lipschitzian real functions, *Proc. Amer. Math. Soc.* **104**, 641–2.

Roehrig, S. and Sine, R. (1981) The structure of the-limit sets of nonexpansive maps, *Proc. Amer. Math. Soc.* **81**, 398–400.

Sadovskii, B. N. (1972) Limit-compact condensing operators, *Russian Math. Surveys* **27**, 85–155.

Schaefer, H. (1974) *Banach Lattices and Positive Operators*, Springer, Berlin, Heidelberg, New York, Tokyo.

Schaffer, J. J. (1976) *Geometry of Spheres in Normed Spaces, Lecture Notes in Pure and Applied Mathematics*, vol. 20, Marcel Dekker, New York, Basel.

Schauder, J. (1930) Der Fixpunktsatz in Funktionalraumen, *Studia Math.* **2**, 171–80.

Schechtman, G. (1982) On commuting families of nonexpansive operators, *Proc. Amer. Math. Soc.* **84**, 373–6.

Schoenberg, J. J. (1953) On a theorem of Kirzbraun and Valentine, *Amer. Math. Monthly* **60**, 620–2.

Siegel, J. (1977). A new proof of Caristi's fixed point theorem. *Proc. Amer. Math. Soc.* **66**, 54–6.

Sine, R. C. (1979) On nonlinear contractions in sup norm spaces, *Nonlinear Anal.* **3**, 885–90.

Sine, R. C. (1987) On the converse of the nonexpansive map fixed point theorem for Hilbert space, *Proc. Amer. Math. Soc.* **100**, 489–90.

Smart, D. R. (1974) *Fixed Point Theorems*, Cambridge University Press, Cambridge.

Smulian, V. (1939) On the principle of inclusion in the space of type (B), *Mat. Sb. (N.S.)*, **5**, 327–8.

Soardi, P. (1979) Existence of fixed points of nonexpansive mappings in certain Banach lattices, *Proc. Amer. Math. Soc.* **73**, 25–9.

Soardi, P. (1982) Schauder bases and fixed points of nonexpansive mappings, *Pacific J. Math.* **101**, 193–8.

Swaminathan, S. (1983) Normal structure in Banach spaces and its generalisations, *Fixed Points and Nonexpansive Mappings* (R. C. Sine, ed.), *Contemporary Mathematics*, vol. 18, American Mathematical Society, Providence, RI, pp. 201–15.

Thele, R. L. (1974) Some results on the radial projection in Banach spaces, *Proc. Amer. Math. Soc.* **42**, 483–6.

Tingley, D. (1984) Noncontractive uniformly lipschitzian semigroups in Hilbert space, *Proc. Amer. Math. Soc.* **92**, 355–61.

Turett, B. (1982) A dual view of a theorem of Baillon, *Nonlinear Analysis and Applications* (S. P. Singh and J. H. Burry, eds.), *Lecture Notes in Pure and Applied Mathematics*, vol. 80, Marcel Dekker, New York, pp. 279–86.

Valentine, F. A. (1943) On the extension of a function so as to preserve a Lipschitz condition, *Bull. Amer. Math. Soc.* **49**, 100–8.

Valentine, F. A. (1945) A Lipschitz condition preserving extension for a vector function, *Amer. J. Math.* **67**, 83–93.

Zizler, V. (1971) On some rotundity and smoothness properties of Banach spaces, *Rozprawy Mat.* **87**.

Index